T0221158

Army Ants

Army Ants

NATURE'S ULTIMATE SOCIAL HUNTERS

Daniel J. C. Kronauer

Harvard University Press

Cambridge, Massachusetts

London, England

2020

First printing

Photographs and illustrations © Daniel J. C. Kronauer, unless credited to other sources

Library of Congress Cataloging-in-Publication Data

Names: Kronauer, Daniel J. C., 1977– author.
Title: Army ants : nature's ultimate social hunters / Daniel J. C. Kronauer.
Description: Cambridge, Massachusetts : Harvard University Press, 2020. | Includes
    bibliographical references and index.
Identifiers: LCCN 2020008537 | ISBN 9780674241558 (cloth)
Subjects: LCSH: Army ants.
Classification: LCC QL568.F7 K76 2020 | DDC 595.79/6—dc23
LC record available at https://lccn.loc.gov/2020008537

For Max, and a world full of wonders

contents

Army Ants

Prologue

The air is hot and humid, and mosquitoes are buzzing around my head. As I walk along a muddy path through the tropical rainforest of La Selva Biological Station in Costa Rica, the giant trees with their enormous buttress roots groan under the heavy load of strangler figs, lianas, epiphytic ferns, and bromeliads (Figure P.1). Besides the humming pests and the sighing trees, everything appears quiet and peaceful. But suddenly a group of birds hop around on the forest floor and in the low vegetation, chirping busily. The birds are so focused on what's going on beneath them that they don't seem to notice as I get closer. They are after a rustling wave of cockroaches, crickets, scorpions, and spiders that scurry through the leaf litter. Yet the birds are not the only ones taking advantage of this curious stampede, as scores of parasitic flies dart down to deposit their eggs or larvae on the unfortunate fugitives.

The insects and spiders have been forced into the open because they are running from what might be the ultimate arthropod nightmare. Like a dark shadow cast on the forest floor, a raiding party of *Eciton burchellii* army ants, hundreds of thousands of individuals strong, is moving across on a daily hunt. The ants form a carpet-like swarm, 10 to 15 meters wide and 2 meters deep, attacking everything that fails to escape. The colonies of social insects, mostly those of other ants, are overrun and their brood chambers plundered. Big arthropods, each vastly larger than the individual army ants, are overpowered by the sheer number of attackers; they are pinned down, stung, and severed into smaller pieces that the ants deem suitable for transport (Figures P.2 and P.3).

FIGURE P.1 A muddy path leads through the tropical lowland rainforest at La Selva Biological Station in Costa Rica. The forest is home to an astonishing number of plant and animal species, including several species of army ants.

FIGURE P.2 Army ants overpower large prey by their sheer numbers. Here, a roach falls victim to an *Eciton burchellii* swarm raid. Alta Floresta, Mato Grosso, Brazil.

FIGURE P.3 Army ants virtually bury large prey items under a mass of workers. In *Eciton burchellii*, a species that hunts on the surface, these clusters often quickly move into the leaf litter, possibly to escape the greedy eyes of swarm-following birds. Here, a swarm raid is engulfing a katydid nymph (*Ischnomela pulchripennis*).

Having observed the battlefield for a few minutes, I follow the ants in the direction in which they carry their loot. Smaller trails of ants coalesce into bigger and bigger columns, until all ants run along a single highway (Figure P.4). The ants carrying prey away from the swarm front are continuously replaced by new warriors that run in the opposite direction to join the battle. The ant highway weaves its way across the rainforest leaf litter, through thicket after thicket, until it finally ends at a fallen tree trunk. And here I find what I came looking for: the army's headquarters. Suspended from the log is an ant bivouac, a sphere approximately half a meter across, made up entirely of the ants themselves (Figure P.5). The scaffolding consists of chains and meshes of thousands of ants that cling together by interlocking little hooks on their feet called tarsal claws (Figure P.6). The prey-laden ants returning from the raid enter this living structure via bridges and tunnels, all formed by the live bodies of their nestmates. This temporary and dynamic formation can be assembled and disassembled as needed, which comes in handy because army ants are nomadic: at dusk, the current bivouac comes down, and the entire colony will move to a new nest site and on to new hunting grounds, over 100 meters across the forest floor.

Even if you haven't had an opportunity to observe army ants firsthand, you've probably heard stories about them. Army ants are popularly known for their ferociousness and unpredictability. In most accounts, they appear suddenly, wreak havoc, and disappear, leaving behind nothing but scorched earth. Harvard entomologist William Morton Wheeler aptly referred to them as the "Huns and Tartars of the insect world" (Wheeler 1910, 246). Wheeler's metaphor was later borrowed by the English evolutionary biologist Sir Julian Huxley, who provided a short yet vivid account of army ant biology in his book *Ants* (Huxley 1930, 80): "These dreaded creatures, with their destructive instincts, their nomadic habits, and the vast number of their hordes, remind us of the invading hosts of Huns or Tartars in our own history. But here again the fundamental differences between man and insect appear. The most restless of the Mongols was capable of settling down to a stable and civilized life. But the driver ants are for ever limited to their way of life by the iron hand of heredity. Their nomadism and

FIGURE P.4 Army ants form columns along which workers travel back and forth between the raid and the bivouac nest. Here, a soldier of *Eciton burchellii* is guarding such a raiding column.

FIGURE P.5 Unlike most other ants, army ants do not build permanent nests but instead form temporary structures, so-called bivouacs, out of their own bodies. Here, an *Eciton burchellii* bivouac, consisting of several hundred thousands of ants, hangs suspended from a fallen tree trunk.

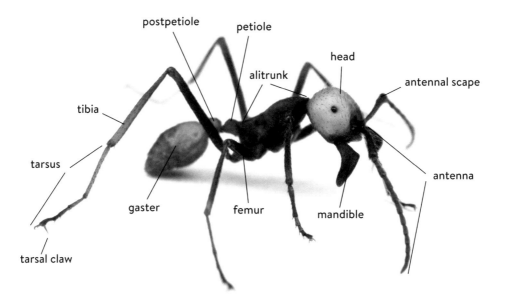

FIGURE P.6 An *Eciton burchellii* submajor worker from La Selva Biological Station in Costa Rica illustrates some of the terms for major morphological features of ants. The ant body is divided into four parts: the head, the alitrunk, the waist, and the gaster. Ant anatomy is somewhat complicated because what technically is the first abdominal segment is fused to the thorax to form the alitrunk. This means that the ant's waist is either composed of the second abdominal segment, the petiole, or in some ants like *Eciton burchellii*, the second and third abdominal segments. In the latter case, the third abdominal segment is called the postpetiole. The antenna of an ant consists of the scape, the pedicel, and the flagellum. The scape is the first antennal segment, and it is connected to the remaining segments by a highly flexible articulation. This allows the ant to fold the pedicel and flagellum up against the scape, giving the appearance of the elbowed antenna characteristic of ants. The pedicel is the second antennal segment, and the flagellum contains all the remaining segments. These so-called flagellomeres can differ in number among different ant species, sexes, and even female castes. Finally, the ant's legs are divided into five parts. The most proximal part (the one attached to the body) is the coxa, followed by the trochanter, femur, tibia, and tarsus. Not all features are separately labeled in the figure.

their ferocity are permanent; their wanderings are barbarous invasions that never end."

The terror of raiding army ants has captured the interest of writers as well. In his short story *The Wild Honey* from 1911, Horacio Quiroga recounts the story of Gabriel Benincasa, who undertakes a trip up the Paraná River to visit his godfather and to experience adventures in the wilderness (Quiroga 1981). Alone in the forest, Benincasa comes across a nest of stingless bees and eats their honey. The wild honey paralyzes poor Benincasa, who then realizes that a swarm of army ants is approaching. When his godfather arrives at the site two days later, he only finds a skeleton, still dressed in the clothes of Benincasa. Quiroga's work has influenced the writings of other Latin American authors, in particular the magic realism of Gabriel García Márquez. His 1967 novel *One Hundred Years of Solitude* concludes with the birth of Aureliano, the last of the Buendía clan. After his mother has died in childbirth and his father has abandoned himself to his grief, the newborn, left unattended, is devoured by a swarm of army ants (García Márquez 1970).

Possibly the most dramatic depiction of army ants can be found in Carl Stephenson's short novel *Leiningen versus the Ants* (Stephenson 1938), which was later adapted into the movie *The Naked Jungle* (1954), starring Charlton Heston as Leiningen. A Brazilian plantation owner, Leiningen, is informed of an approaching army of ants. When he refuses to abandon his plantation, the Brazilian official explains the gravity of the situation: "You're insane! They're not creatures you can fight— they're an elemental—an act of God! Ten miles long, two miles wide— ants, nothing but ants! And every single one of them a fiend from hell; before you can spit three times they'll eat a full-grown buffalo to the bones. I tell you if you don't clear out at once there'll be nothing left of you but a skeleton picked as clean as your own plantation" (98). Although Leiningen thinks himself well prepared for the onslaught, the ensuing battle costs the lives of several plantation workers and nearly his own.

These stories are loosely based on the army ant *Eciton burchellii*. This species, which can be found from central Mexico to Paraguay and southern Brazil, stands out among other Neotropical army ants by the combination of its huge colonies, its broad prey spectrum, its aggressiveness, and its diurnal hunting swarms that cover the surface of the forest floor, cascade into the understory, and surge all the way up into the forest canopy. Most African stories, on the other hand, revolve around a small group of closely related army ant species in the genus *Dorylus,* the so-called driver ants (Figure P.7). These ants have earned their vernacular name because, just as with *Eciton burchellii,* an approaching swarm of ants will drive fleeing insects and other arthropods before its front.

The Swahili term for driver ants is *siafu.* In the poem "Siafu Wameka-zana" ("The Ants Have Mobilized"), Tanzanian poet Saadan Abdu Kandoro uses the symbolism of driver ants attacking a snake as a metaphor for Africans working together to overpower the seemingly invincible British colonists: "Nyoka anababaika, shimoni kwa kujikuna; Siafu wamekazana, nyoka amekasirika" ("The snake raves, in its hole scratching itself. The ants have mobilized, the snake is angry"; Bollag 2001). It is common knowledge across tropical Africa that snakes, metaphorical or not, better beware of driver ants. Ashanti hunters in Ghana, for example, believe that when a python has strangled a large animal it carefully scours the surroundings for driver ants before devouring its prey. This precaution is thought to be necessary because the ensuing feast will leave the snake immobile, unable to escape an approaching ant army (Savage 1847; Cansdale 1961).

The siafu's proverbial bloodthirstiness achieved additional fame in Max Brooks's acclaimed 2006 horror novel about a zombie apocalypse, *World War Z: An Oral History of the Zombie War.* Brooks uses the term "siafu" to refer to the swarming and biting zombies: "The fires, the wreckage . . . the siafu were everywhere. I watched them crash through doors, invade apartments, devour people cowering in corners or on balconies. I watched people leap to their deaths or break their legs and spines. They lay on the pavement, unable to move, wailing in agony as the dead closed in around them" (210).

The legends of army ants killing and devouring humans and other large vertebrates, especially when they are immobilized and helpless, were largely inspired by some of the exaggerated reports of early explorer-naturalists. Paul Du Chaillu, an intrepid explorer and one of the first Westerners to observe gorillas in the wild, traveled in tropical Africa between 1856 and 1859 as part of an expedition organized by the Academy of Natural Sciences at Philadelphia. In his book about his adventures, he provides a particularly gruesome account of driver ants, or "bashikouay," as they were locally known: "When they grow hungry the long file spreads itself through the forest in a front line, and attacks and devours all it overtakes with a fury which is quite irresistible. The elephant and gorilla fly before this attack. The black men run for their lives. Every animal that lives in their line of march is chased. They seem to understand and act upon the tactics of Napoleon, and concentrate, with great speed, their heaviest forces upon the point of attack. In an incredibly short space of time the mouse, or dog, or leopard, or deer is overwhelmed, killed, eaten, and the bare skeleton only remains" (Du Chaillu 1861, 360). Worse still, according to Du Chaillu, the ants had been literally used as executioners by the locals: "The negroes relate that criminals were in former times exposed in the path of the bashikouay ants, as the most cruel manner of putting them to death" (361).

Most of these stories appear far-fetched. As we will see throughout this book, the prey of army ants consists primarily of other invertebrates, and only occasionally do some species take vertebrates, such as small lizards, snakes, or nestling birds. But while tales of their invasions may occasionally be exaggerated, the spectacular social behavior and battle prowess of the army ants cannot be denied. Over the past seventeen years, I have had the privilege to observe army ants in many marvelous places, from the rainforests of Costa Rica, Venezuela, Brazil, Kenya, and Australia, to the Konza Prairie of Kansas, and to study them both in the field and the laboratory. I hope that, with this book, you will join me on a journey of discovery to army ant societies, which resemble large human metropolises in population size but are wondrously different and exotic in all other regards.

The biology of army ants has been the subject of two excellent previous books, Theodore Schneirla's *Army Ants: A Study in Social Organization* from 1971, and William Gotwald's *Army Ants: The Biology of Social Predation* from 1995. However, our understanding of army ant biology has advanced tremendously over the last twenty-five years, fueled by novel insights from fieldwork and, since the early 2000s, the application

FIGURE P.7 Among African army ants, the driver ants in the genus *Dorylus* are most readily encountered because they hunt in the open. In Swahili, they are known as *siafu*. Here, a column of *Dorylus molestus* driver ants traverses the dirt road behind Mount Kenya's Chogoria gate in Kenya.

of molecular genetic tools. In particular, molecular phylogenetic and population genetic work has produced leaps in our understanding of army ant evolution, diet, population structure, reproductive biology, and the many social parasites living inside army ant colonies. Furthermore, the clonal raider ant *Ooceraea biroi* (formerly *Cerapachys biroi*), an emergent model system and close relative of army ants, is finally facilitating mechanistic studies of army ant behavior under controlled experimental conditions in the laboratory. Given these recent breakthroughs, providing an updated synthesis of army ant biology seemed a timely endeavor.

The precise structure and content of the book still required much deliberation, however. On the one hand, I wanted to write a book based on my scientific background and provide a useful reference for my immediate colleagues. On the other hand, I felt that army ants were sufficiently charismatic that it would be a missed opportunity to not also gear the book toward a more general audience with an interest in social insects. I thus decided to attempt the balancing act of addressing both. For example, instead of giving an exhaustive account of army ant biology, I settled on building the book around the life history of two particularly prominent species. The first is the Neotropical swarm raider *Eciton burchellii,* which has been the inspiration for many army ant horror stories. The second is its close relative, the army ant *Eciton hamatum*. Although the overall biology of the two species is similar, they provide an important contrast: *Eciton hamatum* lives in smaller colonies, hunts in columns rather than extensive carpet swarms, and has a much narrower diet, feeding almost exclusively on other social insects. This contrast is useful because it represents two different points along the army ant evolutionary trajectory. However, I extensively draw on other army ant species for context wherever appropriate.

I also chose a specific site, La Selva Biological Station, as the main setting for the book. La Selva's lowland tropical rainforest on the Caribbean slopes of Costa Rica not only is one of the most intensively studied Neotropical ecosystems, but also has been host to much of my recent field research. Because I took most of the photographs in this book at La Selva, it serves as the default locality. In all cases where I took a photograph at a different site, it will be identified in the figure's description. Although this emphasis on La Selva and two particular species meant that I wouldn't cover everything that is known about army ants in a balanced way, I hoped that it would allow me to tell a more engaging story.

In favor of a broader audience, I also decided to illustrate the book mostly with photographs of ants, rather than with original graphs, plots, and tables of scientific data. I explain relevant concepts and approaches as simply as possible and provide an extensive glossary. At the same time, I am retaining the academic style of citing the primary literature in the text, which will provide the specialist readers with quick access to relevant studies. I have attempted to cover the original army ant–specific literature rather exhaustively, but I often refer to recent review articles instead of research papers when discussing general aspects of ant biology. I also have omitted several older studies, especially of army ant morphology, anatomy, and taxonomy, because they are either outdated or purely descriptive. Those studies are reviewed in detail in Gotwald's 1995 book. Finally, interwoven in this book is also my personal story in the form of field anecdotes from the tropical rainforests of the world. This approach was of course a challenge. Nevertheless, I hope that, with all this in mind and irrespective of your background, you will embrace the compromises made in the final product.

I will begin the book at the dawn of the eighteenth century, and detail the discovery of my two main protagonists, *Eciton burchellii* and *Eciton hamatum,* by early European naturalists as they explored the Neotropics. Chapter 1 also tells the story of how a lost type specimen resulted in great taxonomic confusion around these two species. In Chapter 2, I reconstruct the evolutionary history of army ants, and outline how their conspicuous life history was derived from that of their more cryptic ancestors. This chapter places *Eciton burchellii* and *Eciton hamatum* in the broader context of army ant biodiversity. Army ants are defined by three functionally and evolutionarily interconnected traits: hunting in massive groups, frequently relocating the nest, and multiplying by splitting large colonies. These three traits are discussed in detail in Chapters

3, 4, and 5, respectively, again with a focus on my two lead characters. Finally, Chapter 6 will discuss the myriad of organisms that infiltrate and exploit army ant colonies as social parasites, with the community of parasites associated with *Eciton* army ants at La Selva Biological Station taking center stage. In the opening chapter, the life history of Neotropical leaf-cutting ants of the genus *Atta* will serve as a useful contrast to army ant biology, being more representative of ants in general. Similar to army ants, leaf-cutting ants live in very large and sophisticated colonies, and, as we will see in later chapters, these two types of gargantuan societies frequently interact, usually peacefully but sometimes in open combat.

CHAPTER ONE

Discovery of the Main Protagonists, and the Mystery of a Lost Type

Early Descriptions

The first European mention of *Eciton burchellii,* the infamous swarm raider, comes from an anonymous letter dated January 24, 1701, received from Paramaribo in the Dutch colony of Suriname, and read by a Monsieur Homberg to the Royal Academy of Sciences in Paris (Anonymous 1701). Even though the author of the letter is unknown, the text resembles a description that was published shortly after by the German naturalist and artist Maria Sibylla Merian to such an extent that sheer coincidence seems unlikely. Merian had been fascinated by the diversity of insects from an early age, and the wondrous metamorphosis of insects had received her special attention. Not only was she a meticulous observer, but also an excellent illustrator of plants and animals. In 1699, Merian traveled to Suriname, where she spent twenty-one months in a religious community close to Paramaribo. Her sojourn culminated in her beautifully illustrated opus magnum from 1705, a book on the metamorphosis of the insects of Suriname, *Metamorphosis Insectorum Surinamensium* (Merian 1705).

Plate 18 of Merian's *Metamorphosis* depicts what seems to be an amalgamation of *Eciton burchellii* army ants and *Atta* leaf-cutting ants on a guava tree (Figure 1.1). In leaf-cutting ant manner, the ants have partly defoliated the tree and are harvesting additional leaves. But they are also attacking spiders and a roach, displaying typical army ant behavior. The

FIGURE 1.1 Plate 18 from Maria Sibylla Merian's book *Metamorphosis Insectorum Surinamensium* shows the interactions between different species on the branch of a guava tree (*Psidium guineense*) in Suriname (Merian 1705). Leaf-cutting ants (*Atta cephalotes*) are harvesting leaves while army ants (*Eciton burchellii*) are attacking spiders and a roach. A pinktoe tarantula (*Avicularia avicularia*) is preying on an ant, while another one has caught a ruby-topaz hummingbird (*Chrysolampis mosquitus*) on its nest. All plates in the original 1705 edition of *Metamorphosis* were produced from engravings based on Merian's watercolors. Image courtesy of the Linda Hall Library of Science, Engineering & Technology.

only Neotropical army ant that will ascend higher up into the vegetation to attack a wide range of arthropods is *Eciton burchellii,* so it is safe to assume that Merian's depiction is loosely based on the biology of that species.

In the accompanying text, Merian describes how ants in South America can defoliate an entire tree in a single night, carrying the leaves back to their nest. This account of leaf-cutting ant behavior is immediately followed by a description of army ant behavior: "When they want to go somewhere where no path is leading, they build a bridge. To do so, the first ant lies down and bites into the wood, the second attaches by clinging to the first, likewise the third to the second, the fourth to the third, and so on. Then they drift in the wind until they get blown to the other side. Then all the thousands walk across like over a bridge. These ants have an eternal feud with the spiders and all insects of the country. Once every year they emerge in huge numbers from their cellars. They enter the houses and go from one room to the next, sucking all the animals dry, big and small. In no time have they devoured a big spider, because the ants descend on the spider in great numbers, so the spider cannot save itself. They also go from one room to the next, so that the humans have to retreat as well. Once the ants have devoured everything inside the house, they move on to the next, and finally back into their cellar" (Merian 1705, 18; my translation).

Eciton burchellii army ants do in fact build bridges with their own bodies, even though the formation of such a bridge occurs by a process that is rather different from the one described by Merian (see Chapter 3) (Figures 1.2 and 1.3). Merian also accurately described how army ants enter and clean out human habitations by killing and removing the vermin they encounter, although she incorrectly interpreted this as a seasonal phenomenon.

It might seem surprising that Merian failed to clearly distinguish between leaf-cutting ants and army ants in her painting. As far as ants go, the two are about as different as it gets behaviorally, and taxonomically speaking they even belong to different subfamilies. However, it has to be kept in mind that, at the time, insects were still regarded by many as evil

FIGURE 1.2 Army ants are the masters of living modular architecture, the ant version of LEGO toys. This image shows *Eciton burchellii* workers sticking together to bridge a gap in the leaf litter. While the ants forming the bridge can remain motionless for hours, others can now travel at full speed across the crevice, treading comfortably on the backs of their uncomplaining sisters.

FIGURE 1.3 A close-up photograph of an *Eciton burchellii* bridge showing how the ants adhere to the vegetation and each other with their tarsal claws.

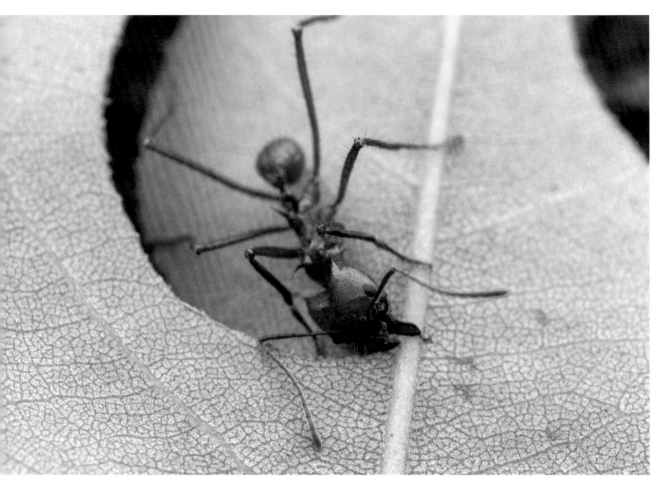

FIGURE 1.4 Leaf-cutting ants are serious agricultural pests in the American tropics because of their appetite for human crops. Here, a worker of *Atta cephalotes* is doing what she does best: cutting a leaf.

beasts of the devil that were born from mud in a process called spontaneous generation. Accordingly, tropical ants had not been studied scientifically, let alone placed in a rigorous classification system. The first formal description of a leaf-cutting ant, *Atta cephalotes,* for example, was provided by the great Swedish biologist and "father of taxonomy" Carl Linnaeus, half a century after the publication of *Metamorphosis,* with direct reference to Merian's original account (Linnaeus 1758) (Figures 1.4 and 1.5).

FIGURE 1.5 Leaf-cutting ants use freshly cut leaves as a substrate to cultivate their main food source: the basidiomycete fungus *Leucoagaricus gongylophorus*. This image shows the queen and workers of a young *Atta cephalotes* colony on their incipient fungus garden. Mature colonies of this species can number over a million ants and occupy a vast network of fungus chambers, each one the size of a human head.

Arguably the first European naturalist who passed down more extensive notes on New World army ants was the Spaniard José Celestino Mutis (Wilson and Gómez Durán 2010). At age twenty-eight, Mutis was appointed private physician to the viceroy of New Granada. New Granada was a vast and largely unexplored land that included today's Colombia, Ecuador, Guyana, Venezuela, and Panama. Mutis set sail from the Spanish port of Cádiz and set foot in the Caribbean coastal town of Cartagena de Indias in 1760. Although Mutis's foremost passion was botany, he had corresponded with Linnaeus, who had encouraged him specifically to collect and study ants (Wilson and Gómez Durán 2010).

For example, Mutis made the effort to track army ants through the thicket, until he was rewarded with an unforgettable sight, an army ant bivouac: "The ants had not begun to excavate a nest in the ground but instead had settled on the sides of the log that touched the ground and along the upper part of the roof of the log, so that in the middle it seemed like a cluster hung in the air. . . . The entire external surface of this nest, from the only side where it could be seen, was so symmetrical and smooth, and also so perpendicular, that it seemed to have been leveled. I never tired of inspecting this beautiful spectacle" (Wilson and Gómez Durán 2010, 42–49).

Mutis also described the ants' nomadic lifestyle in some detail and followed several colony emigrations to the new bivouac site. He even estimated the size of an entire army ant colony, based on the dimensions of the bivouac and the traffic flow in colony emigrations; he came to the staggering conclusion that there must be approximately 3 million workers in a colony. Even though later authors have arrived at slightly lower estimates for the size range of *Eciton burchellii* colonies—approximately 500,000 to 2 million workers according to Schneirla (1971), and up to about 650,000 workers according to Franks (1985)—army ant societies are clearly gigantic. Or, as Mutis put it, "The numbers may seem incredible except for somebody who has seen the colonies" (Wilson and Gómez Durán 2010, 73).

As Europe's political power and reach in the world grew, more explorers returned with specimens of seemingly innumerable new species from faraway lands. Linnaeus, who had made it his life's work to catalogue this immense diversity, was faced with the impossibility of achieving this goal single-handedly. He therefore encouraged Johann Christian Fabricius, another prolific student of his, to take on an especially species-rich group, the insects. Fabricius did so with great vigor, describing and naming nearly 10,000 species and establishing the field of systematic entomology.

Fabricius rendered the first formal description of an *Eciton* army ant as part of a longer treatise on general insect taxonomy, his *Species Insectorum* (1782). The workers of *Eciton* army ants are highly polymorphic. In other words, they come in a variety of different shapes and sizes. Although this polymorphism is somewhat continuous, the worker force of an *Eciton burchellii* colony, for example, can be divided into three castes: regular workers, submajors, and majors (or soldiers) (Powell and Franks 2006; see also Franks 1985). The smallest regular workers are only about 4 millimeters long whereas the soldiers measure three times that in body length and, unlike any of the other worker castes, sport impressively exaggerated mandibles (Figure 1.6). The workers of *Eciton hamatum* are similarly polymorphic (Figure 1.7). Fabricius's description was based on an *Eciton* soldier, and he named the corresponding species *Formica hamata. Formica* simply means "ant" in Latin, and Fabricius initially described all kinds of different ants as part of this one genus. The species name *hamata* stems from the Latin *hamatus,* which means "hook shaped" and refers to the mandibles of the *Eciton* soldier. At the time, taxonomic descriptions were usually brief and not necessarily supplemented with drawings of the specimens in question.

The first visual depiction of the *Formica hamata* type specimen was provided by the French entomologist Pierre André Latreille in his 1802 book *Histoire Naturelle des Fourmis* (*Natural History of the Ants*). Latreille, who had trained as a Roman Catholic priest, was incarcerated in 1793 under the threat of being executed because he failed to swear an oath of allegiance to the state, which had become a requirement for priests during the French Revolution. But Latreille knew how to put his knowledge of entomology to good use. He impressed the prison's doctor by finding and recognizing a rare beetle in the dungeons. The doctor sent the beetle to a local naturalist, who not only confirmed Latreille's identification but was also familiar with his work and ultimately was able to get Latreille released. The species that saved Latreille's life, *Necrobia ruficollis,* the red-necked bacon beetle, had been described by none other than Fabricius; once Latreille had regained his freedom, the two entomologists engaged in regular correspondence.

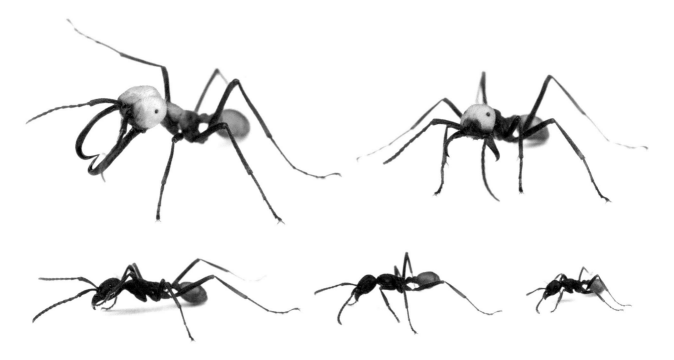

FIGURE 1.6 The workers of an *Eciton burchellii* colony come in different sizes, shapes, and even colors. The smallest workers are approximately 4 millimeters in body length, and the largest workers measure approximately 12 millimeters (Schneirla 1971). Although the workers fall along a morphological continuum, three discrete worker castes have been identified via morphometric analyses: soldiers (or majors; upper left), submajors (upper right), and regular workers (bottom) (Powell and Franks 2006). Regular workers make up the vast majority of a colony's population, while soldiers and submajors each account for only a couple of percent (Schneirla 1971). Images of all workers are to scale.

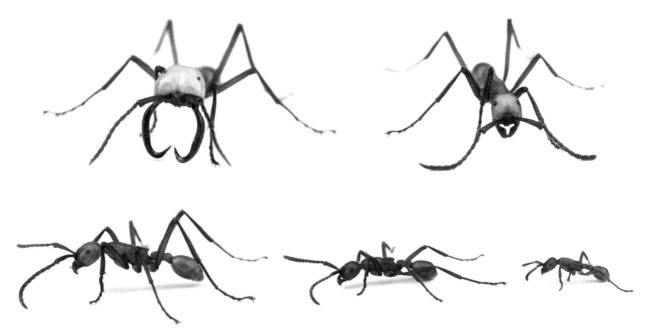

FIGURE 1.7 Workers of *Eciton hamatum* match those of *Eciton burchellii* in terms of their polymorphism and size range. As in *Eciton burchellii*, the soldiers can be easily identified by their ivory-colored heads and immense, hook-shaped mandibles. *Eciton* workers are not only polymorphic but also differ in behavior. The very smallest workers usually remain inside the bivouac and take care of the eggs and young larvae, subma-jors are specialized for carrying bulky items, and soldiers defend the colony (e.g., Schneirla 1971; Topoff 1971; Franks 1985; Powell and Franks 2005). Images of all workers are to scale.

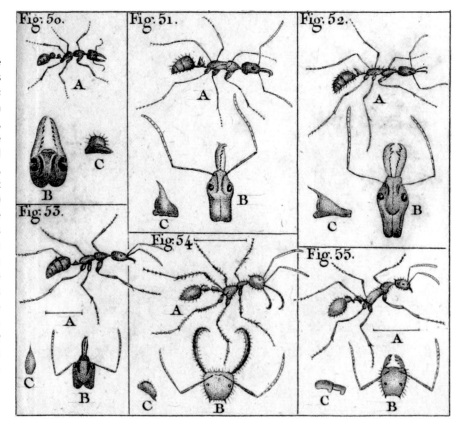

FIGURE 1.8 This part of plate VIII in Pierre André Latreille's book *Histoire Naturelle des Fourmis* (1802) depicts a number of different ants, among them Fabricius's type specimen of what he called *Formica hamata* (figure 54, bottom center). The adjacent figure 55 (bottom right) shows an ant that Latreille described separately as *Formica curvidentata*, which was later recognized as a different worker caste of the species described by Fabricius (Smith 1858). Image from the Biodiversity Heritage Library. Contributed by the University Library, University of Illinois at Urbana-Champaign.

The drawings of Fabricius's *Formica hamata* type appear next to drawings of another ant, which Latreille described as *"Formica curvidentata,"* and which he placed in a different group of ants (Figure 1.8). *Formica curvidentata,* however, was later synonymized with the Fabricius type—that is, the two were recognized as belonging to the same species (Smith 1858). From today's perspective, it seems obvious that the *Formica curvidentata* specimen is an *Eciton* worker, but simply not a soldier worker like the Fabricius type. However, it has to be kept in mind that neither Fabricius nor Latreille had ever seen a colony of *Eciton* army ants, and their descriptions were entirely based on isolated workers from a species with extreme worker polymorphism. Finally, arguably the most permanent of Latreille's contributions to the study of *Eciton* army ants was that, in 1804, he established the very genus *Eciton,* to which he assigned Fabricius's *Formica hamata* (Latreille 1804). So by 1804 the species *Eciton hamatum* was taxonomically official.

Unfortunately, Fabricius's type specimen eventually went missing. The Franciscan missionary Thomas Borgmeier, who would provide what is still the most recent taxonomic revision of New World army ants in his 1955 treatise *Die Wanderameisen der neotropischen Region* (*The Army Ants of the Neotropical Region*) (Borgmeier 1955; see also Borgmeier 1953, 1958), surmised that the specimen might be deposited at the museum of the University of Kiel in northern Germany, where part of the Fabricius collection is housed. A substantial part, however, is stored at the Natural History Museum of Denmark in Copenhagen, where Fabricius had been a professor before his later appointment in Kiel.

I had the pleasure of seeing this collection firsthand. While I was working on my doctoral thesis at the University of Copenhagen, Edith Rodríguez, then an ant taxonomist at the Central University of Venezuela, and I one day decided to check out the ant collection at the Natural History Museum. As we looked through the trays of pinned ants, we couldn't believe our eyes. There she was, among Fabricius's type specimens, a single, beautifully preserved *Eciton* soldier on a separate pin with one of Fabricius's original labels that had but a single word written on it: "hamata." However, to our great surprise, the specimen in question did not belong to *Eciton hamatum* as it is known today, but to *Eciton burchellii,* the infamous swarm raider. Even though the two species are closely related and superficially similar in overall appearance, they can be easily distinguished in the field, both morphologically and in terms of their biology. With about 150,000 to 500,000 workers at most, the colonies of *Eciton hamatum* are considerably smaller than those of *Eciton burchellii.* Rather than attacking a broad range of prey species in massive swarms, their raids are organized in discrete columns that target almost exclusively the colonies of other ants and social wasps (Rettenmeyer 1963a; Schneirla 1971; Gotwald 1995; Powell 2011) (see Chapter 3). To understand the root of this confusion, we have to turn to the explorers and taxonomists who studied army ants in the nineteenth century.

Understanding the Origins of Biological Species

Throughout the eighteenth century, the book of Genesis provided what seemed to be a satisfactory explanation for the diversity of life on Earth. Rather than seeking a deeper understanding of the origin of biological species, Linnaeus and Fabricius had regarded it as their task to bring order to God's creation. Or, as Linnaeus himself put it, "God created— Linnaeus arranged." In the first half of the nineteenth century, however, a new idea was in the air, an idea that would forever transform the way we think about the relationships between the living creatures of the world and our own place among them. This simple yet powerful notion proposed that biological species had not been created by God and were immutable, but rather that they changed slowly over time in a process that Charles Darwin so famously described as evolution by natural selection. This ultimately led to the realization that all organisms on Earth are related. Every species, including humans, army ants, and Merian's guava tree, can be traced back to a single origin of life. What's more, at any point in time, every living organism has spent the exact same amount of evolutionary time diverging from this last universal common ancestor of all living things, currently somewhere around 3.5 billion years. Consequently, there is no particular hierarchy among extant species, no *scala naturae* that naturally places God above humans and humans above all other organisms.

This idea did not sit well with many of Darwin's contemporaries, who gained much comfort from the superior role that God had assigned them among His creation. Once this concept had been internalized, however, overwhelming evidence for its validity could be found virtually everywhere. The tropics, with their immense biodiversity and bizarre forms of animals and plants, seemed like an especially intense battleground for survival and natural selection. Darwin's account of his own travels in South America therefore soon inspired others to follow (Darwin 1839).

Having read and avidly discussed Darwin's early work, two friends from Leicester, England, embarked on a ship headed to the Brazilian city of Belém (then Pará) in 1848. Their main goal was to collect insects and other exotic specimens, and to accumulate data pertaining to the then hotly debated question of the origin of species and the process of evolution. After a year of collecting, the friends decided to split up in order to make their efforts as efficient as possible. One of them, Alfred Russell Wallace, who later became known for independently deducing the theory of evolution by natural selection, explored along the Rio Negro (Wallace 1878). The other, Henry Walter Bates, focused on the Rio Solimões. When Wallace returned to England in 1852, the cargo on his ship caught fire, and most of his painstakingly accumulated collection was destroyed. In 1859, after having spent eleven years in the Amazon, Henry Walter Bates finally returned to England as well. Aware of Wallace's misfortune, he had shipped his immense collection of specimens of over 14,000 species distributed across three separate ships back to England. Among those specimens, a specialist for Hymenoptera at the British Museum named Frederick Smith recognized ten species of army ants, eight of which he described as new to science.

Bates provided a detailed account of Neotropical army ants in his 1863 book *The Naturalist on the River Amazons:* "When the pedestrian falls in with a train of these ants, the first signal given him is a twittering and restless movement of small flocks of plain-coloured birds (ant-thrushes) in the jungle. If this be disregarded until he advances a few steps farther, he is sure to fall into trouble, and find himself suddenly attacked by numbers of the ferocious little creatures. They swarm up his legs with incredible rapidity, each one driving his pincer-like jaws into his skin, and with the purchase thus obtained, doubling in its tail, and stinging with all its might. There is no course left but to run for it; if he is accompanied by natives they will be sure to give the alarm, crying 'Tauoca!' and scampering at full speed to the other end of the column of ants. The tenacious insects who have secured themselves to his legs then have to be plucked off one by one, a task which is generally not accomplished without pulling them in twain, and leaving heads and jaws sticking in the wounds" (760).

When Bates published his travel memoirs in 1863, Frederick Smith's formal taxonomic descriptions of the army ants he had encountered had already become available (Smith 1855, 1858, 1860), and Bates was able to refer to the different species by their proper scientific names. Intriguingly, however, Bates used the name *Eciton hamatum,* which today denotes a species with intermediate colony sizes that hunts in columns and feeds almost exclusively on other social insects, when furnishing a description that can be unambiguously attributed to *Eciton burchellii,* the great swarm raider. The flocks of birds he mentions only gather at *Eciton burchellii* swarm raids, and his description of the ants' diet also uniquely fits that of *Eciton burchellii:* "Wherever they move, the whole animal world is set in commotion, and every creature tries to get out of their way. But it is especially the various tribes of wingless insects that have cause for fear, such as heavy-bodied spiders, ants of other species, maggots, caterpillars, larvae of cockroaches and so forth, all of which live under fallen leaves, or in decaying wood" (760). It thus appears that, at the time, the name *Eciton hamatum* was associated with the species that today is known as *Eciton burchellii.* But if that is the case, then when did things get mixed up?

The confusion can be traced back to a paper by the Austrian myrmecologist Gustav Mayr from 1886, which is partly based on a reexamination of the type specimens of Frederick Smith, who had died seven years earlier. In this paper, Mayr synonymizes some of Smith's specimens with Fabricius's *Formica hamata* type as *Eciton hamatum,* and furnishes a redescription that fits the species that from then onward would be known as *Eciton hamatum,* the column raider. There seems to be no evidence that Mayr had actually examined Fabricius's original type specimen when he came to this unlikely conclusion. After all, it implied that Bates's collecting efforts had omitted what might be the most formidable ant of all, the swarm raider *Eciton burchellii.* However, this misguided move put Mayr in a position to once again describe *Eciton burchellii* as a species new to science. He did so in honor of his fellow myrmecologist Auguste Forel, naming the species *Eciton foreli* (Mayr 1886).

Forel was arguably one of the most eccentric and prominent characters in myrmecology at the time. In 1869, at the age of twenty-one, Forel had published his first paper on the predatory behavior of the fire ant *Solenopsis fugax,* in which he criticized Mayr's work (Forel 1869). "I sent my paper on *Solenopsis fugax,* among other things, to the classifier of ants, Dr. Gustav Mayr, in Vienna, whom I regarded with some disfavour, for in his works he had referred in rather contemptuous terms to my idol, Huber [Pierre Huber, whose 1810 book *Recherches sur les Moeurs des Fourmis Indigènes* was regarded by Forel as "the Bible of myrmecology"], while he himself had been guilty of blunders in the treatment of biological problems, since he was anything but a biologist. In my paper I had defended Huber. Mayr was annoyed, and replied in an offended tone" (Forel 1937, 62). The young Forel apologized in the hope to avoid a dispute with the well-known scholar, and over the years they developed what seems to have been an overall amicable relationship. However, one cannot escape the irony of the fact that Mayr committed a major taxonomic blunder quite literally in Forel's name.

Unfortunately, the new classification stuck, and later authors followed Mayr in their usage of the two scientific names. So we have seen how the original *Eciton hamatum* acquired the name *Eciton foreli,* and how the name *Eciton hamatum* became incorrectly reassigned. This leaves the question of how *Eciton foreli* shed Forel's name and became ultimately known as *Eciton burchellii.* This question turns out to be of interest beyond nomenclature because it leads us to a general challenge in army ant biology: how do you associate the winged males of a given species with their female counterparts? Although we now have molecular genetic tools available that make the challenge less daunting, this problem remains prevalent in army ant taxonomy to this day (e.g., Berghoff et al. 2003a; Ward 2007; Kronauer et al. 2007c; Schöning et al. 2008a).

Discovery of the Males

The principle of priority in nomenclature posits that the correct scientific name for a species is the oldest applicable name, but surely, when Mayr described his *Eciton foreli* in 1886, there was no ant species known by the name of *Eciton burchellii.* So how did this name gain priority?

From 1825 to 1830, about two decades before Bates and Wallace embarked on their expedition to the Amazon, another English explorer and naturalist, William John Burchell, was traveling in Brazil. Burchell collected insects that were flying at night and were attracted to lights.

FIGURE 1.9 The *Eciton burchellii* male was originally described as *Labidus burchellii*, based on a specimen collected by the English explorer-naturalist William John Burchell in Brazil. Because the large, winged, and unusual looking army ant males are most readily collected at lights during the night rather than directly from army ant colonies, their taxonomic association long remained a mystery.

Many nocturnal insects employ transverse orientation, in other words, they use a distant source of light as a reference point to which they keep a fixed angle while flying, not unlike sailors have used the North Star for navigation. This form of orientation evolved long before the times of man-made noncelestial light sources, and, instead of using the moon as a proper point of reference, moths and many other insects will circle around a lightbulb or find their demise in the flame of a candle. Among Burchell's nocturnal collection was a series of peculiar insects that looked like large male wasps, and similar insects had previously been collected at lights in tropical Asia and Africa. The African representatives in particular were so beefy that they became known as "sausage flies."

These and other specimens collected by Burchell were sent to the Oxford University Museum and were described in 1842 by the English entomologist John Obadiah Westwood as belonging to the genus *Labidus* (Westwood 1842). These descriptions were part of a larger monograph of what Westwood called the "Dorylides," which included the Asian and African species, all of which were solely known from males at the time. Westwood surmised that, in fact, they were male ants, and he designated one of these rare and unusual insects as the type specimen for *Labidus burchellii* (Figure 1.9).

We know today that in most cases all the ants you will find in a colony are female: the queen and her workers. Males usually do not partake in the social life of an ant colony and are produced only seasonally in many species. Their adult life often does not last longer than a few days, and after a brief tenure inside their natal nest, they disperse on the wing to mate before they die. This also applies to army ant males; unsurprisingly then, for the longest time nobody had ever encountered males inside an army ant colony. But nobody had ever dared to take apart an *Eciton* colony to look inside either. For obvious reasons, as Henry Walter Bates pointed out: "I have observed its legions in processions of great extent, but up to the present time I have been unable to meet with the other sexes; their societies are so numerous and the sting of the insects so severe, that an attack on one of their colonies is not to be rashly undertaken" (cited in Smith 1858, 149). Consequently, following Fabricius's initial description of an *Eciton* soldier in 1782, it took an entire century until a male was finally found inside a colony and described.

Wilhelm Müller was visiting his brother, the famous German biologist Fritz Müller who had emigrated to Brazil in 1852, in Blumenau, when, on February 28, 1885, he encountered a trail of *Eciton* army ants in Fritz's garden. Wilhelm decided to follow the colony's nightly emigrations (Müller 1886). During the emigration on March 1, Wilhelm noticed a single *Labidus* male that had lost his wings and seemed to be guided along the path by the workers, prompting Wilhelm to conclude that this male must have belonged to the colony. Upon his return to Europe, Wilhelm gave the ants he had collected in Blumenau to no other than Auguste Forel, who described them in a short taxonomic paper (Forel 1886). While Forel identified the wingless male as *Labidus burchellii,* he concluded that the corresponding workers belonged to Fabricius's *Eciton hamatum.* This was the first clear demonstration that some of the bizarre wasp-like insects that had been sporadically collected from lights at night were actually male *Eciton* army ants.

So what were the consequences of this finding? Because the species name *hamatum* (Fabricius 1782) is older than, and therefore has priority over, the species name *burchellii* (Westwood 1842)—and, similarly, the genus name *Eciton* (Latreille 1804) has priority over the genus name

Labidus (Jurine 1807)—the species should have retained the name *Eciton hamatum.* However, Gustav Mayr's confusing revision, which was published in the same year as the papers by Müller and Forel, led later authors to believe that Forel had incorrectly identified Müller's ants (e.g., Wheeler 1921, Borgmeier 1955). It was thus concluded that the ants instead belonged to Forel's namesake species, Gustav Mayr's "ghost species" *Eciton foreli* from 1886. Accordingly, Carlo Emery, another of the great myrmecologists at the time, soon thereafter formally synonymized Westwood's *Labidus burchellii* with Mayr's *Eciton foreli* as *Eciton burchellii* (Emery 1896; see also Emery 1890).

The convoluted story of how the magnificent swarm raider *Eciton burchellii* ended up with the incorrect scientific name illustrates the importance of type specimens. When these exemplars are destroyed, lost, or otherwise become inaccessible, it can be extremely difficult for scientists to decide what species exactly previously published research refers to. It also makes the general point that taxonomic reference collections are pivotal in advancing our understanding of biodiversity and life history, and that they serve as historical records of biological exploration.

Although the army ant workers and males were finally matched up, the most important colony member still remained to be discovered.

Discovery of the Queens

The ants you see scavenging for dead insects on the pavement, the ants that raid the sugar bowl in your kitchen, and the stinging ants that ruin your picnic party—all of them are worker ants. The same is true for the many thousands of ants that constitute an army ant raid. In fact, nearly all the ants you will ever encounter will be workers. In the vast majority of cases, however, the female workers cannot mate with males. They are also often sterile, or at least have greatly reduced reproductive potential. Instead, ants have a second female caste that is entirely specialized for egg laying and is usually much larger than the workers: the queens (see Figure 1.5).

The best chance of seeing queen ants is during the time of mating flights. In many ants, large colonies produce new queens and males during a certain time of the year. The young queens and males have wings, and upon some environmental trigger, the colonies of a given population release them to join the annual mass orgy. In leaf-cutting ants, for example, the mating flights typically occur after the first heavy downpours of the rainy season. Once mating has concluded, the males have fulfilled their purpose and die. The young queens, on the other hand, land on the ground, shed their wings, and each digs for herself a little chamber into the ground where she will start to lay eggs. This is the founding stage of a new colony.

For a long time—several years in some species—all these eggs will develop into workers. Over time, the workers will expand the little chamber into a full-sized ant nest, they will forage for food and defend the colony, and, most importantly, they will raise more and more new workers. This is the ergonomic stage of colony growth. Once the colony has reached a certain size, it will start to produce new males and queens, which will once again disperse and go on a mating flight. The colony has now reached the reproductive stage. The single queen that initially founded the colony will be the mother of all the workers, males, and young queens the colony will ever produce. Ant queens clearly are very special insects.

This description of a basic ant colony life cycle fits species like the leaf-cutting ant *Atta cephalotes* fairly well, but it is of course highly generalized. In fact, nearly any variation on this theme you can imagine has evolved in one ant species or another. Army ants deviate in two important ways. First, only the males leave the colony to disperse, while the queens remain inside the colony at all times. Second, the colonies of other ants can produce thousands of young males and queens at a time; army ant colonies produce many males but never more than half a dozen or so queens. Army ant queens are therefore among the rarest of all insects. This, together with the previously mentioned practical difficulties associated with casually sorting through a colony of a gazillion irate army ants, made the search for the army ant queen a considerable challenge indeed.

One of the first persistent attempts to discover the enigmatic queen of *Eciton burchellii* was that of Wilhelm Müller while studying army ants in his brother's garden in Blumenau. The sudden appearance of eggs in the colony led Müller to conclude that a queen must have been present, which spurred him on to follow the colony over the course of two and a half weeks. After he had repeatedly smoked out the ants from their abode inside a hollow tree trunk, to no avail, he finally resorted to destroying the nest altogether. Despite all these brute efforts, and to Müller's great frustration, Her Majesty refused to grant him an audience.

It was not until well into the twentieth century that the mystery of the *Eciton* queen was finally solved. William Morton Wheeler was spending the summer of 1920 at the Tropical Laboratory of the New York Zoological Society at Kartabo in British Guiana together with his son Ralph. The field station had been established only two years earlier by William Beebe, a naturalist, explorer, and popular science writer from New York, who had recently written two articles about his own encounters with army ants (Beebe 1917, 1919). Colonies of *Eciton burchellii* were particularly common around the area, and Ralph eventually found a colony that had bivouacked inside the base of a hollow tree only a few hundred meters away from the station buildings—a convenient location to pester the queen and her army.

Wheeler set out to once again smoke out an *Eciton* colony, but not without taking the necessary precautions: "We ringed the legs of two chairs with carbolated vaseline, planted them in front of the opening, crouched on their seats and with long tweezers placed in the bottom of the cavity a lot of moist bamboo leaves and paper. A match was applied and soon a dense smudge filled the cavity and even issued from cracks in the old wood at a height of nearly twenty feet from the ground. The ants remained quiet for some time, but when the smoke grew denser decided to move, and columns of workers and soldiers began to emerge from the top of the long orifice, crawled out over the bark and descended to the ground" (Wheeler 1921, 296–298).

It turned out that Wheeler was incredibly lucky. The colony he had chosen happened to harbor one of those rare broods of a few young queens and several thousand males. The males were still enclosed in large silken cocoons, 20 millimeters in length, which were covered and slowly propelled down the evacuation route by large masses of workers. However, in two of those halting masses, Wheeler discovered young queens: "The two females had evidently recently emerged for their colors were very brilliant and the delicate golden pile on their bodies was intact" (Wheeler 1921, 300). In other words, these were not the mother queens of the colony, but two princesses, sisters of the males, whose aspirations to inherit the army ant throne were now to be drowned in formalin, a common fixative in entomological research at the time.

Measuring 21 and 23 millimeters, the two queens were even larger than the males, and, unlike their brothers and the young queens of most other ant species, they did not bear wings (Figures 1.10 and 1.11). This was an exciting discovery, because a lack of wings in young ant queens goes hand in hand with a rather unusual mode of colony reproduction, in which a large mature colony splits in two, and each of the daughter colonies is headed again by a single queen. This mode of colony reproduction, known as "colony fission," is strikingly different from the typical ant life cycle described earlier in that colonies never go through a founding phase. Colony fission is very rare among ants and social insects in general, but, as discussed in more detail in Chapters 2 and 5, it is the standard mode of colony reproduction among army ants and has important implications for their biology.

FIGURE 1.10 The queen of *Eciton burchellii* has royal status and is usually covered by a dense mass of protective worker ants. For this portrait she was temporarily removed from her colony's emigration column.

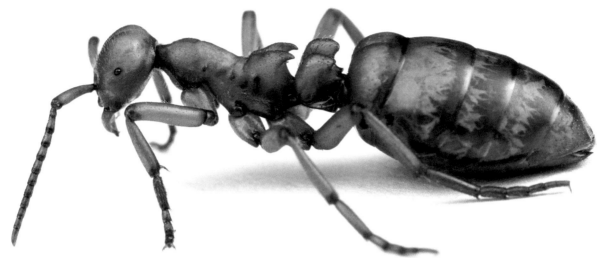

FIGURE 1.11 This image shows the queen of *Eciton hamatum*, briefly removed from her colony for a photo session. The single queen is the mother of all the workers in an army ant colony. Army ant queens are arguably among the most elusive insects. William Morton Wheeler, who had previously discovered the queen of *Eciton burchellii*, discovered the queen of *Eciton hamatum* in the summer of 1924 on Barro Colorado Island in the Panama Canal Zone (Wheeler 1925). La Selva Biological Station, Costa Rica.

The discovery of the *Eciton* queens by Wheeler shall conclude our account of over two centuries of early exploration and research on *Eciton* army ants. But the plundering hordes of ants had of course roamed tropical forests around the world long before Maria Sibylla Merian or any other human being ever laid eyes on them. The next chapter will therefore take us back not to the Age of Enlightenment, but to the age of dinosaurs, the mid-Cretaceous, over 100 million years ago. This is when ants first evolved from their wasp-like ancestors. I will discuss where on the ant family tree army ant behavior has subsequently originated, and how it functionally relates to the biology of army ants' closest relatives.

CHAPTER TWO

Army Ant Ancestry

Exactly how old ants are is a bit of a contentious issue. The first insects clearly identifiable as ants are fossils entrapped in amber from France and Myanmar, dating back to the middle of the Cretaceous period, about 100 million years ago (LaPolla et al. 2013; Barden 2017). Many of these specimens belong to groups that have long gone extinct, but some resemble extant ants, those still alive today. This implies that ants originated during the height of the dinosaurs. In fact, there are roughly as many known fossil ant species as there are species of dinosaurs (Barden 2017). This time estimate based on the fossil record agrees fairly well with current estimates based on DNA, the text-like molecule that constitutes the blueprint for almost all organisms.

Because the sequence of DNA "text" changes over evolutionary time at a more or less constant rate, sequence differences between extant species can be used to gauge the time that has elapsed since they started to diverge from their common ancestor. This feature is known as the molecular clock. Although molecular clock estimates have varied quite a bit (Brady et al. 2006; Moreau et al. 2006; Moreau and Bell 2013), the latest study places the most recent common ancestor of all extant ants somewhere between 103 and 124 million years ago, only slightly older than the oldest fossils found (Borowiec et al. 2019). This does not seem unreasonable, especially keeping in mind that the fossil record is incomplete, and ants older than 100 million years might either not have been preserved as fossils or remain to be discovered (Ward 2014; Barden 2017).

In Cretaceous fossil records, ants are always relatively rare, never exceeding 1.5 percent of insect inclusions (Barden 2017). By the Eocene at around 50 million years ago, however, ants account for approximately 20 percent of insect fossils in some localities; in approximately 19 million years old Dominican amber from the Miocene, the prevalence of ants has risen to around 30 percent (Wilson and Hölldobler 2005; Ward 2014; Barden 2017). By then ants had achieved ecological dominance, which they have maintained to this day.

Our planet is currently populated by well over 14,000 different species of ants, and more than 400 ant species have been documented at La Selva Biological Station alone (Longino et al. 2002). Furthermore, it has been estimated that ants make up at least one-third of the total insect biomass on Earth (Wilson 1990; Wilson and Hölldobler 2005). A truly whopping amount of ants! Over the many millions of years of their existence, ants have evolved an astonishing variety of life histories. Maybe not surprisingly given their diversity and numbers, ants are indispensable in terrestrial ecosystems. They act as scavengers, aerate the soil, disperse seeds, harvest and prune vegetation, and form tight-knit symbioses with plants and other insects. And then there are army ants, the top arthropod predators in tropical rainforests around the world. This ecological dominance of ants is intimately linked to a remarkable innovation: the evolution of eusociality.

To be classified as eusocial, a species must fulfill three requirements. First, there must be reproductive division of labor, where some individuals, usually the queens, lay eggs, while others, the workers, forego reproduction and instead perform tasks such as brood care, nest maintenance, foraging, and colony defense. Second, overlapping generations must live together; and third, adults must cooperate to care for the young (Wilson 1971). Cooperation and division of labor are thought to make eusocial societies much more efficient and productive than the sum of their individual parts. A solitary organism faces trade-offs between different functions such as reproduction and foraging, and thus is constrained to remain a jack-of-all-trades and a master of none. These trade-offs, however, disappear as insect societies become more and more integrated over evolutionary time. Ultimately, a eusocial colony can afford to produce individuals that are highly specialized for specific tasks—such as egg-laying, foraging, or defense—but perform terribly at others.

At this point, from an evolutionary perspective, the ant colony can be conceptualized as a single "organism" or "superorganism," in which natural selection acts largely at the level of the social group (Wheeler 1911; Hölldobler and Wilson 2009; Boomsma and Gawne 2018). This evolutionary progression has allowed for the extreme range of morphological differences in the female castes of *Eciton* army ants and *Atta* leaf-cutting ants, for example (see Figures 1.5–1.7, 1.10, and 1.11). As a solitary insect an *Eciton* queen or soldier would be doomed, but as part of the colony their respective performance at reproduction and defense is pretty much unparalleled. Further, while most solitary organisms are not particularly good at multitasking and have to carry out different functions sequentially, an ant colony can lay eggs, forage, feed the larvae, and defend the nest, all at the same time. However, competition over food and reproduction in solitary organisms pose substantial hurdles that must be overcome in order to evolve eusociality (Korb and Heinze 2016).

Accordingly, this extreme form of social living occurs only sporadically across the tree of life. Yet a good portion of the world's eusocial taxa are found within the order Hymenoptera. These include the ants, some bees such as honeybees and bumblebees, and some wasps such as yellow jackets and paper wasps. Ant, wasp, and bee societies have many similarities, among them the fact that workers are exclusively female. The main reason for these similarities is that hymenopteran societies have evolved along similar evolutionary trajectories from subsocial ancestors that performed extended brood care (Wilson 1971). This "brood care route" to eusociality required that the first helpers that remained with the mother were able to care for the young, a behavior that, in hymenopterans, is only displayed by females (Queller and Strassmann 1998; Korb and Heinze 2008; Kronauer and Libbrecht 2018).

At the same time, ant societies differ from those of other eusocial hymenopterans in important ways. Possibly the most striking feature that sets them apart is that ant queens shed their wings after the nuptial flight and that ant workers are always wingless, unlike worker bees and wasps. It is thought that, unimpeded by bulky and delicate wings, ants became particularly apt to nest and forage in the soil and other tight spaces, giving the first ants access to rich resources inaccessible to other eusocial hymenopterans and further boosting their ecological success (Lucky et al. 2013). Additionally, with the loss of wings in the worker caste, another evolutionary constraint had arguably been removed: ant workers no longer needed to remain aerodynamic and able to fly. This may have facilitated evolution of the exaggerated queen–worker dimorphism and worker subcaste diversity observed in some ants (Peeters and Ito 2015).

Ant Diversity

The over 14,000 ant species alive today are grouped into seventeen different subfamilies (Figure 2.1). Each subfamily constitutes a major phylogenetic clade, an evolutionary entity that contains a common ancestor and all of its descendants. Evolution has produced enormous diversity in general appearance and life history among ants, but a few broad trends emerge. Because these will be important in informing the discussion of repeated evolutionary origins of army ant–like behavior across the ants later in this chapter, I will provide a brief overview of ant evolution and biodiversity in general. This background will also be useful in Chapter 3, which deals with ants as army ant prey.

Perhaps the most enigmatic ant subfamily is the Martialinae, which contains a single species, *Martialis heureka. Martialis* had eluded ant enthusiasts for centuries. Finally, in the twilight of May 9, 2003, a single worker emerged from the soil of the primary rainforest near Manaus in Brazil, was immediately recognized as "unusual," and collected by the sharp-sighted myrmecologist Christian Rabeling. This remarkable find of a 2 millimeter long ant in the vastness of the Amazon should soon have the scientific community abuzz with excitement. Not only did it constitute a new subfamily, but it also appeared to be an ancient relic of ant family history and only distantly related to most other extant ants (Rabeling et al. 2008). It is now believed that the Martialinae, together with another cryptic subfamily, the Leptanillinae, indeed form a sister group to all other living ants (Borowiec et al. 2019) (see Figure 2.1). This implies that these two subfamilies in particular can provide insights into the biology of the first modern ants.

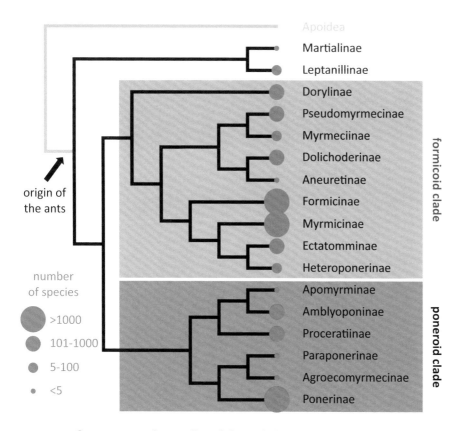

FIGURE 2.1 Our current understanding of the evolutionary relationships between the seventeen extant ant subfamilies (based on Borowiec et al. 2019). The subfamilies Leptanillinae and Martialinae are thought to have diverged early during ant evolution from the remaining subfamilies, which are divided into two major clades, the formicoids and the poneroids. The vast majority of army ants belong to the subfamily Dorylinae, and a few additional army ants are found in the ponerine genus *Leptogenys*. However, species that resemble army ants in their biology to different degrees have evolved in several additional subfamilies (see Table 2.1). The approximate number of species in each ant subfamily is indicated by the size of the green disks. Within the Aculeata (the stinging wasps), the ants are believed to be most closely related to a group called the Apoidea, which includes the bees and apoid wasps (Branstetter et al. 2017a).

Table 2.1. Known life history characteristics of select species with different degrees of army ant-like behavior outside of the subfamily Dorylinae.

Subfamily	Species	Raid initiation	Prey spectrum	Nomadism	Mode of colony founding[1]	
Amblyoponinae	*Onychomyrmex hedleyi*	Spontaneous	Mostly centipedes	Yes	Probably dependent	
Leptanillinae	*Leptanilla japonica*	?	Centipedes	Yes	Probably dependent	
Myrmicinae	*Carebara diversa*	Spontaneous	Diverse arthropods, other animals, and plant matter	Yes	Independent	
Myrmicinae	*Carebara silenus*	Spontaneous	Diverse arthropods and other animals	Yes	Independent	
Myrmicinae	*Tetramorium uelense*	Scouts	Termites	?	Probably independent	
Myrmicinae	*Pheidole titanis*	Scouts	Termites	?	?	
Ponerinae	*Leptogenys chinensis*	Scouts	Termites	Yes	Probably dependent	
Ponerinae	*Leptogenys cyanicatena*	Spontaneous	Millipedes	Yes	Probably dependent	
Ponerinae	*Leptogenys diminuta*	Scouts	Diverse arthropods	Probably	Probably dependent	
Ponerinae	*Leptogenys nitida*	Spontaneous	Diverse arthropods	Yes	Probably dependent	
Ponerinae	*Leptogenys distinguenda*	Spontaneous	Diverse arthropods and other animals	Yes	Probably dependent	
Ponerinae	*Megaponera analis*	Scouts	Termites	Yes	Dependent	
Ponerinae	*Neoponera commutata; Neoponera laevigata; Neoponera marginata*	Scouts	Termites	Yes	Probably independent; mostly independent, and occasionally dependent in *N. marginata*	
Ponerinae	*Simopelta oculata; Simopelta pergandei*	?	Ants	Yes	Probably dependent	

Note: Question marks indicate missing information.

[1] Ant queens can found new colonies either independently by themselves, or dependently with the assistance of workers from the mother colony in the case of army ants and species with similar life histories. Queens in the former case are usually winged (alate), and queens in the latter case are usually permanently wingless (ergatoid). In this table, "independent" and "dependent" colony founding refers to species for which the process has been observed directly. "Probably independent" and "probably dependent" refer to species with winged and wingless queens, respectively, but direct observations are lacking.

[2] Phasic colony cycles in which developmentally synchronized cohorts of larvae suppress egg-laying are common in dorylines (see Tables 2.2 and 2.3),

Permanently wingless queens	Reproductive queens per colony	Workers per colony	Phasic colony cycles[2]	References
Yes	One	850 (±341 SD)	Probably not	Miyata et al. 2003, 2009
Yes	One	100–200	Probably not	Masuko 1989, 1990, 2019
No	Often one, sometimes several	Ca. 250,000	No	Moffett 1984, 1988b, 1988c, 2010, 2019
No	Often one, sometimes several	Over 100,000	No	Moffett 1988a, 1988c, 2010, 2019
No	?	?	?	Longhurst et al. 1979b; AntWeb.org
?	?	?	?	Feener 1988
Yes	Unclear	200–300	No	Maschwitz and Schönegge 1983
Yes	One	500–1,000	No	Peeters and De Greef 2015; Arimoto and Yamane 2018
Yes	One	60–439	No	Maschwitz and Mühlenberg 1975; Attygalle et al. 1988; Maschwitz and Steghaus-Kovac 1991; Ito and Ohkawara 2000
Yes	One	200–1,000	No	Duncan and Crewe 1994
Yes	One	Ca. 50,000	No	Maschwitz et al. 1989; Witte and Maschwitz 2000, 2002
Yes	One	320–856; median 1,039	No	Longhurst and Howse 1979; Villet 1990; Hölldobler et al. 1994b; Bayliss and Fielding 2002; Frank 2020; Erik Frank, personal communication
No	One in *N. commutata*; often several in *N. marginata*	Up to 880 in *N. commutata*; 561–1,581 in *N. marginata*; up to at least 1,700 in *N. laevigata*	No	Hermann 1968; Hölldobler and Traniello 1980; Mill 1982, 1984; Leal and Oliveira 1995; Hölldobler et al. 1996; Schmidt and Overal 2009
Yes	One	Ca. 1,000	Possibly	Gotwald and Brown 1966; Mackay and Mackay 2008; Kronauer 2009; Kronauer et al. 2011a

but have not been unambiguously demonstrated in species outside of the Dorylinae. For example, the developmentally synchronized broods observed in colonies of *Leptanilla japonica* and *Onychomyrmex hedleyi* rather seem to be a seasonal phenomenon (Masuko 1990; Miyata 2003, 2009). Developmentally synchronized broods have also been found in colonies of *Simopelta* (Gotwald and Brown 1966; Kronauer et al. 2011a). Given that *Simopelta* is tropical and therefore likely not seasonal, this suggests that colonies might indeed undergo phasic reproductive cycles similar to those of many dorylines (Gotwald and Brown 1966; Kronauer et al. 2011a). In cases where non-synchronous broods have been reported (i.e., eggs, larvae, and pupae occurring simultaneously inside a colony), the species was deemed to not be phasic.

The long front legs and thin, forceps-like mandibles of *Martialis* suggest that it might be a specialized predator of soft-bodied arthropods (Rabeling et al. 2008). To this day, however, the species is known from only three workers, and no behavioral observations of live specimens exist (Brandão et al. 2010). Similarly, the Leptanillinae hunt in the soil as specialized predators, and the life history of some species strikingly resembles that of army ants (Table 2.1). Taken together, this opens the possibility that, just like most army ants, the first modern ants were specialized subterranean predators (Rabeling et al. 2008; Lucky et al. 2013; Lanan 2014). However, unlike army ants, these species probably lived in small colonies.

The remaining ants can be classified into two major clades, the poneroids and the formicoids (see Figure 2.1). With a few exceptions, poneroids are still predatory, live in the soil, and have relatively small colonies. With about 1,200 known species, the Ponerinae constitutes the largest poneroid subfamily, and, as we will see, some ponerines have evolved army ant lifestyles (Schmidt and Shattuck 2014) (Figure 2.2; see Table 2.1). A medium-sized poneroid subfamily with somewhat more than 100 species is the Amblyoponinae, which contains the Dracula ants (Figure 2.3). The name is derived from the ants' habit of piercing the cuticle of their larvae to suck hemolymph, the insect equivalent of blood. Just like some leptanillines and ponerines, some amblyoponines have evolved army ant–like behavior (see Table 2.1). The bullet ant *Paraponera clavata,* known for its potent sting, likewise belongs to the poneroids and is the sole representative of the subfamily Paraponerinae (Figure 2.4).

Within the formicoids, the Dorylinae are sister to the remaining subfamilies (see Figure 2.1). This subfamily contains the vast majority of army ants, including my two main protagonists, *Eciton burchellii* and *Eciton hamatum,* and I will discuss it separately in the next section. In fact, all dorylines are predatory, and most are still entirely subterranean. In contrast, many of the remaining formicoids are frequently observed above ground, and many form associations with plants. Some feed on nectar or the honeydew produced by sap-sucking homopterans, while

FIGURE 2.2 The ponerine genus *Leptogenys* contains species with a wide range of foraging strategies. This image shows *Leptogenys* workers (probably *Leptogenys diminuta*) at the front of a group raid. This particular raiding party was several dozen workers strong, and retrieved a large earwig, among other things. Iron Range National Park, Australia.

FIGURE 2.3 Ants in the subfamily Amblyoponinae are predators, and many are specialized on centipedes, such as the Dracula ant *Stigmatomma pallipes*, shown here with a larva. Walden Pond, Massachusetts.

FIGURE 2.4 A worker of the bullet ant *Paraponera clavata*, arguably the most venomous ant in the world, roams the forest floor at La Selva Biological Station, Costa Rica. As the name suggests, being stung by a bullet ant feels like being shot.

others are obligate herbivores or engage in intricate ant–plant mutualisms (Wilson and Hölldobler 2005; Moreau et al. 2006; Lanan 2014). Even though this part of the ant tree of life contains the great majority of species, army ant–like lifestyles are exceedingly rare, and restricted to a handful of species in the subfamily Myrmicinae such as the marauder ants of the genus *Carebara* (Moffett 1984, 1988a, 1988b, 2010, 2019) (Figure 2.5; see Table 2.1). With about 6,800 species, the Myrmicinae (from the Greek word *myrmex,* meaning "ant") is by far the largest ant subfamily and encompasses a broad range of ant lifestyles, including leaf-cutting ant agriculture (see Figures 1.4 and 1.5).

Second in line, with approximately 3,200 species, is the subfamily Formicinae. The Formicinae include the well-known weaver ants (genus *Oecophylla*) and carpenter ants (genus *Camponotus*) (Figures 2.6 and 2.7). Despite the large size of this subfamily, there are no known formicines that show army ant–like predatory behavior. However, a different lifestyle that is centered on plundering the colonies of other ant species has evolved repeatedly in the Formicinae, and it is in this subfamily where it has been taken to the highest levels of sophistication: slave-making. The colonies of some formicine slave-makers, like *Polyergus rufescens,* send out individual scouts that then recruit large raiding parties once they have encountered a host colony (e.g., Mori et al. 2001). As we will see, this behavior bears striking resemblance to scout-initiated group raids of army ant relatives. In other species, like *Formica subintegra,* colonies send out larger exploring parties that leave the nest in a column and eventually fan out. Once a host colony has been detected, some of the explorers return to the nest to recruit reinforcement (Talbot and Kennedy 1940) (Figure 2.8). However, the purpose of these raids is quite different. Unlike army ants, which simply kill and consume their victims, the slave-makers raise the captured brood into adults. Seemingly unaware of their roots, the enslaved ants then perform foraging and nest-maintenance tasks, contributing to the prosperity of the slave-maker colony.

FIGURE 2.5 Some species in the genus *Carebara* (subfamily Myrmicinae) show army ant-like behavior by forming large raids that overwhelm other arthropods. Workers in this genus can show extreme size polymorphism, as is shown here for *Carebara affinis*. Iron Range National Park, Australia.

FIGURE 2.6 Weaver ants (here *Oecophylla smaragdina*) form their nests out of leaves, which they weave together with silk produced by their larvae. A single colony can occupy many such nests across several trees. Iron Range National Park, Australia.

FIGURE 2.7 Weaver ants are very aggressive and attack, kill, and consume most arthropods that wander into their territory. Here, several *Oecophylla smaragdina* weaver ants take on a large *Camponotus confusus* worker. Iron Range National Park, Australia.

FIGURE 2.8 The raids of slave-making ants serve to restock the slave ant population in the nest. This picture shows a queen of the slave-maker *Formica subintegra* surrounded by two of her daughters on the right, as well as a slave ant of the species *Formica subsericea* from the same nest on the left. Colonies of this species send out exploration parties of several dozen workers, and additional workers are recruited once a host nest has been located. *Formica subintegra* colonies also frequently relocate by moving into the plundered host nests (Talbot and Kennedy 1940). The Rockefeller University Field Research Center, Millbrook, New York.

Slave-making in ants is a fascinating form of social parasitism, and, unlike army ants, slave-makers usually raid the colonies of closely related host species whose overall biology is sufficiently similar to facilitate the social integration of the slave workers (Hölldobler and Wilson 1990). Evidently, raiding has evolved repeatedly across the ant phylogeny as a strategy to overwhelm the well-defended colonies of other social insects.

Some of the most intricate interactions between ants and plants are found in the formicoid subfamilies Dolichoderinae and Pseudomyrmecinae, including species in the dolichoderine genus *Azteca* that form mutualistic associations with *Cecropia* trees. These intimate relationships were studied by Wilhelm Müller's brother Fritz. The trees house the ants inside their hollow stems, while the ants defend the trees against herbivores. However, the trees are so fond of their aggressive little tenants that they not only provide shelter but even produce nutritious food bodies, the appropriately named Müllerian bodies, to feed the ants (Figures 2.9 and 2.10). Similarly, *Pseudomyrmex* needle ants (subfamily Pseudomyrmecinae) inhabit the swollen thorns of bullhorn acacias. As in the *Azteca–Cecropia* mutualism, the ants defend the acacias against herbivores and overgrowth by competing plants. The acacias also provide nutrition in the form of sugary secretions from extrafloral nectaries, along with food bodies that are displayed at the tips of the leaflets and harvested by the ants (Figures 2.11 and 2.12). The acacia food bodies are called Beltian bodies, after the English mining engineer and naturalist Thomas Belt, a contemporary of Fritz Müller who first described the association between ants and plants in his 1874 classic *The Naturalist in Nicaragua*. As we will see in Chapter 3, Belt was also a diligent observer of army ant behavior.

FIGURE 2.9 Several species in the genus *Azteca* (subfamily Dolichoderinae) inhabit the hollow stems of *Cecropia* trees. The tree produces glycogen-rich and proteinaceous Müllerian food bodies for the ants on the brown pad, or trichilium, at the base of the leaf petiole visible in the background. The food bodies can rarely be seen on the trichilia of trees with active ant colonies, such as this one, because they are immediately harvested by the trees' assiduous tenants. Here, *Azteca* spec. (probably *Azteca xanthochroa*) ants are patrolling the area around an entrance hole to their nest inside a *Cecropia* spec. (probably *Cecropia obtusifolia*) tree trunk.

FIGURE 2.10 A look inside the tree's hollow stem reveals the *Azteca* ants' nursery and pantry: the two larger white items, including the one carried by the ant, are larvae, while the smaller, egg-shaped white items are Müllerian food bodies. Probably *Azteca xanthochroa* and *Cecropia obtusifolia.*

FIGURE 2.11 Many species in the genus *Pseudomyrmex*, the needle ants of the subfamily Pseudomyrmecinae, form tight associations with ant plants. Arguably the most famous example, which was described by the English naturalist Thomas Belt, are Central American *Pseudomyrmex* species that inhabit bullhorn acacias. Like in the mutualism between *Cecropia* trees and *Azteca* ants, the acacias provide both board and lodging. This image shows two *Pseudomyrmex* spec. (probably *Pseudomyrmex spinicola*) ants drinking from an extrafloral nectary of their *Vachellia* spec. host plant, while additional workers in the background are retrieving yellow Beltian food bodies that the acacias produce at the leaf tips. Carara National Park, Costa Rica.

FIGURE 2.12 Bullhorn acacias get their name from the shape of their paired thorns. The thorns are hollow, and provide living space for *Pseudomyrmex* ants that defend the tree against herbivores and competing plants. Here, a *Pseudomyrmex* spec. (probably *Pseudomyrmex spinicola*) worker is seen at the entrance to such a domatium on a *Vachellia* spec. acacia. Carara National Park, Costa Rica.

The Subfamily Dorylinae

When people talk about "army ants," they usually refer to species in the subfamily Dorylinae, including the Neotropical *Eciton burchellii* and *Eciton hamatum,* as well as the African *Dorylus* driver ants. About twenty different doryline army ants occur at La Selva alone, and this diversity seems representative of Neotropical rainforests in general (Longino et al. 2002; O'Donnell et al. 2007). Some of them hunt during the day, while others are nocturnal. Some form large raids in the forest leaf litter, and in some cases they can ascend all the way up into the canopy of rainforest tree giants. Others are almost entirely subterranean and hardly ever seen above ground. Some have a broad prey spectrum that includes a large variety of arthropods and sometimes even plant matter. Others are specialized predators of a few related species of ants (Rettenmeyer et al. 1983; Hoenle et al. 2019). The ecological impact of such diverse communities is arguably enormous. In the remainder of this chapter, I will discuss the evolutionary trajectory that gave rise to the lifestyle and impressive diversity of army ants.

The oldest known doryline fossil hails from Baltic amber and is approximately 44 million years old (Barden 2017). This is consistent with two recent studies estimating that, based on molecular clock analyses, the subfamily most likely arose sometime around 74 or 55 million years ago, respectively (both estimates come with considerable uncertainty and large confidence intervals; Borowiec 2019; Borowiec et al. 2019). The biology of most of the approximately 700 doryline species is poorly known, but all the species that have been studied in any detail show at least some behavioral resemblance to army ants (Table 2.2).

The phylogenetic relationships between different doryline taxa have long been plagued with ambiguity, which made earlier evolutionary interpretations error prone (Gotwald 1979, 1995). These earlier reconstructions were solely based on morphological characters, and advances in DNA sequencing techniques have since led to a revolution in our understanding of army ant taxonomy and phylogeny (Brady 2003; Brady and Ward 2005; Brady et al. 2014; Borowiec 2016, 2019). The amount of differences in DNA sequences between different organisms not only

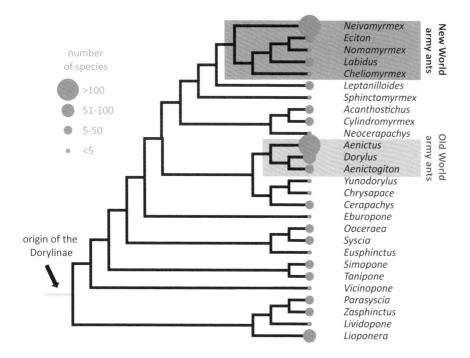

FIGURE 2.13 The genus-level phylogeny of the ant subfamily Dorylinae, which contains most of the army ants (based on Borowiec 2019). Doryline army ants are distributed across two clades, the Old World army ants and the New World army ants. According to the most recent phylogenetic estimate, these two clades have evolved independently from non-army ant dorylines. The green disks indicate the approximate number of species in each genus.

allows us to estimate how long ago they shared a common ancestor, but also the branching pattern between different evolutionary lineages, which is usually depicted as a phylogenetic tree. Given that modern DNA sequencing data sets are very large, they can often resolve even tricky cases with high confidence.

The dorylines are currently classified into twenty-seven genera (Borowiec 2016, 2019) (Figure 2.13). Intriguingly, most species are subterranean predators, possibly not unlike the most recent common ancestor of modern ants (Lucky et al. 2013). And they are not just predators. The vast majority are predators of other social insects, mostly ants (Tables 2.2 and 2.3). In fact, it has been proposed that dorylines were the first group of ants that became specialized on an ant diet. In a way, the

Table 2.2. Known life history characteristics of select species in the subfamily Dorylinae other than the army ants.

Species	Raid initiation	Prey spectrum	Nomadism	Permanently wingless queens	
Cerapachys cf. sulcinodis	?	Ants; occasionally other arthropods	?	Yes	
Cylindromyrmex whymperi	Scouts	Termites	Probably	No	
Leptanilloides erinys	?	?	Probably	Yes	
Lioponera cf. turneri	Scouts	Ants	Probably	Yes	
Lioponera daikoku	?	Ants	?	No	
Lioponera suscitata	?	Ants	?	No	
Ooceraea biroi	Scouts	Ants	Yes	Queenless	
Parasyscia opaca	?	Ants	?	?	
Parasyscia sp. 1	?	Ants	?	?	
Syscia augustae	?	?	Yes	Yes (brachypterous)	
Syscia humicola	?	?	?	Yes	
Syscia minuta	?	?	?	No	
Yunodorylus eguchii	?	Unclear, but found in association with termites	?	Yes	
Zasphinctus cf. steinheili	?	Ants	Probably	Yes	

Note: Question marks indicate missing information. Colony founding has not been directly observed in any of the species listed here, but it seems likely that colonies are founded dependently in species with permanently wingless queens, and possibly independently in species with winged queens. Given that *Ooceraea biroi* colonies are queenless and workers are totipotent and reproduce asexually, new colonies probably arise haphazardly when colony fragments split off.

[1] Phasic colony cycles have not been directly observed in this species, but collected colonies contained developmentally synchronized broods.

	Reproductive queens per colony	Workers per colony	Phasic colony cycles	References
	Multiple	Up to 2,000	Yes	Mizuno et al. 2019; Riou Mizuno and Fuminori Ito, personal communication
	One	Ca. 100; possibly more	?	De Andrade 1998; Gobin et al. 2001
	One	Ca. 100	Possibly[1]	Borowiec and Longino 2011
	One or two	Ca. 100–300	No	Hölldobler 1982; Bert Hölldobler, personal communication
	Often one, sometimes several	Up to 44	No	Idogawa and Dobata 2018
	One	Up to 40	No	Ito et al. 2018
	All workers can reproduce asexually	Ca. 150–600; ca. 250–750	Yes	Tsuji and Yamauchi 1995; Ravary and Jaisson 2002, 2004; Ravary et al. 2006; Chandra et al. n.d.
	One	Ca. 75–100	Possibly[1]	Wilson 1958b
	One	13	Yes	Ito et al. 2018
	One	29	Possibly[1]	Wheeler 1903; Borowiec 2016; Longino and Branstetter n.d.
	One or two, rarely more	Up to 21	?	Masuko 2006
	One	Ca. 30	Possibly[1]	Longino and Branstetter n.d.
	One	Several thousand	Yes	Eguchi et al. 2016; Riou Mizuno and Fuminori Ito, personal communication
	Multiple	Ca. 80–400	Yes	Buschinger et al. 1989

Table 2.3. Known life history characteristics of select species of army ants in the subfamily Dorylinae.

Species	Prey spectrum	Surface active[1]	Degree of worker polymorphism[2]	Reproductive queens per colony	
OLD WORLD ARMY ANTS					
Aenictus laeviceps	Ants; occasionally other arthropods	Yes	Low	One	
Dorylus laevigatus	Diverse arthropods, other animals, and plant matter	No	High	One	
Dorylus wilverthi	Diverse arthropods; occasionally other animals and plant matter	Intermediate	High	One	
NEW WORLD ARMY ANTS					
Eciton burchellii	Ants; diverse arthropods and other animals	Yes	High	One	
Eciton hamatum	Mostly ants and social wasps	Yes	High	One	
Labidus coecus	Diverse arthropods, other animals, and plant matter	No	High	One	
Labidus praedator	Diverse arthropods, other animals, and plant matter	Intermediate	High	One	
Neivamyrmex kiowapache	?	No	Low	Three to thirteen	
Neivamyrmex nigrescens	Ants and termites	Intermediate	Low	One	
Neivamyrmex pilosus	Ants; occasionally other arthropods	No	Low	One	
Nomamyrmex esenbeckii	Ants; occasionally other arthropods	No	Intermediate	One	

Note: Question marks indicate missing information. All species listed in this table are army ants with mass raiding, nomadism, and dependent colony founding with permanently wingless queens. Even though raid initiation has only been described for a few species, it can be assumed that raids in all army ants begin spontaneously, and recruitment occurs predominantly at the raid front. Only a small subset of relatively well-studied species is listed to illustrate some of the life history diversity.

[1] Army ants range from species that are rarely if ever observed above ground (hypogaeic; "no" in this table; here this includes also species like *Neivamyrmex pilosus* and *Nomamyrmex esenbeckii* that sometimes briefly raid above ground), to those that nest underground but are regularly encountered raiding above ground ("intermediate" in this table), to those that nest and raid above ground (epigaeic; "yes" in this table).

[2] Although many army ant species have essentially monomorphic workers, some display among the most extreme levels of worker polymorphism among the ants. In this table, species are categorized as having "low" worker polymorphism if the ratio between the body length

Workers per colony	Phasic colony cycles	Key references
60,000–110,000	Yes	Chapman 1964; Schneirla and Reyes 1966, 1969; Schneirla 1971; Topoff 1971; Rościszewski and Maschwitz 1994; Gotwald 1995; Hirosawa et al. 2000
325,000	No	Berghoff et al. 2002a, 2002b, 2003a, 2003b
20,000,000–22,000,000	No	Raignier et al. 1974; Gotwald 1995; Schöning et al. 2008b

Workers per colony	Phasic colony cycles	Key references
500,000–2,000,000; 300,000–650,000[3]	Yes	Rettenmeyer 1963a; Schneirla 1971; Franks 1985; Gotwald 1995
150,000–500,000; 48,000–125,000[4]	Yes	Rettenmeyer 1963a; Schneirla 1971; Topoff 1971; Gotwald 1995
?	No	Rettenmeyer 1963a; Rettenmeyer and Watkins 1978; Gotwald 1995; Powell and Baker 2008
Over 1,000,000; close to 4,000,000[5]	Ambiguous	Rettenmeyer 1963a; Rettenmeyer and Watkins 1978; Fowler 1979; Gotwald 1995; Powell and Baker 2008; Baudier and O'Donnell 2016
Up to 50,000	Ambiguous	Borgmeier 1955; Rettenmeyer and Watkins 1978; Mackay and Mackay 2002; Kronauer and Boomsma 2007b; Snelling and Snelling 2007
80,000–140,000; 9,500–100,000[6]	Yes	Schneirla 1958, 1971; Rettenmeyer 1963a; Topoff 1971; Mirenda et al. 1980; Topoff and Mirenda 1980a; Topoff et al. 1980a; Gotwald 1995
?	Possibly[7]	Borgmeier 1955; Rettenmeyer 1963a; Rettenmeyer and Watkins 1978; Gotwald 1995; Powell and Baker 2008
Ca. 500,000–1,200,000[8]	Possibly[7]	Borgmeier 1955; Rettenmeyer 1963a; Rettenmeyer and Watkins 1978; Gotwald 1995; Powell and Baker 2008

of the largest to the smallest workers is 2:1 or below, and as having "high" polymorphism if the ratio is 3:1 or above. Ratios in between are classified as "intermediate."

[3] The two alternative colony size estimates for *Eciton burchellii* are from Schneirla (1971) and Franks (1985), respectively.

[4] The two alternative colony size estimates for *Eciton hamatum* are from Schneirla (1971) and Powell (2011), respectively.

[5] The two alternative colony size estimates for *Labidus praedator* are from Rettenmeyer (1963a) and Fowler (1979), respectively.

[6] The two alternative colony size estimates for *Neivamyrmex nigrescens* are from Schneirla (1958, 1971) and Mirenda et al. (1980), respectively.

[7] Phasic colony cycles have not been directly observed in this species, but collected colonies contained developmentally synchronized broods.

[8] This is based on the rough estimate of Rettenmeyer (1963a) that colonies of *Nomamyrmex esenbeckii* are "between three and eight times the size" of *Eciton hamatum* colonies.

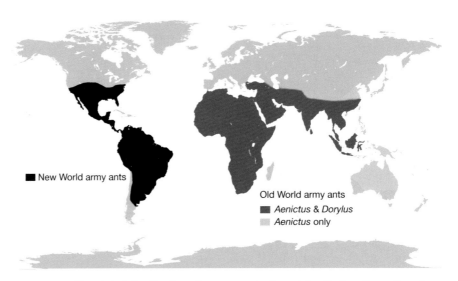

FIGURE 2.14 The global distribution of army ants in the subfamily Dorylinae. Doryline army ants fall into two clades, the New World and the Old World army ants. The New World army ants include the genera *Cheliomyrmex*, *Eciton*, *Labidus*, *Neivamyrmex*, and *Nomamyrmex*. The Old World army ants include the genera *Aenictus* and *Dorylus*, as well as the enigmatic genus *Aenictogiton*. The distribution of *Aenictogiton* is not indicated here, but it is contained within the distribution of *Aenictus* and *Dorylus* in continental Africa. The shown distributions are based on data from Borowiec (2016).

dorylines "invented" specialized myrmecophagy. As ants in general were becoming more and more abundant and ecologically dominant around the time that dorylines first appeared, this evolutionary innovation proved highly successful, arguably leading to rapid initial diversification in this lineage (Borowiec 2016, 2019).

Among the dorylines, army ants make up two clades, the Old World army ants and the New World army ants (see Figure 2.13). The Old World army ants inhabit Africa and Asia, and even reach into Oceania; the New World army ants are found in the Americas (Figure 2.14). Over the past decades, there has been a wavering debate over whether Old World and New World army ants are descended from a single army ant ancestor, or whether they have evolved independently from non–army ant dorylines with small colonies. This is an important question because independent evolutionary origins of seemingly associated, army ant–characteristic traits would further support the conjecture that such characteristics are

FIGURE 2.15 While doryline army ants have huge colonies and highly derived queens that are much larger than the workers, the colonies of other dorylines are considerably smaller, and the size of their queens in most cases is much less exaggerated. Here, a queen of *Syscia madrensis*, a non–army ant doryline, is shown on the left next to her worker daughters. This species occurs in the southern United States, but it is rarely encountered (Longino and Branstetter, n.d.). This experimental colony was collected in Arizona's Chiricahua Mountains, and maintained and photographed at the Rockefeller University.

in fact functionally related. Although this debate is likely to continue, the most recent phylogenetic study indeed favors the hypothesis of at least two independent evolutionary origins of army ants within the Dorylinae (Borowiec 2019) (see Figure 2.13).

Based on biogeographic reconstructions, it seems likely that the Dorylinae originated in the Old World, possibly in the African tropics. This is also where the Old World army ants would eventually arise and subsequently spread into Asia on several independent occasions, and ultimately to Australia. At very roughly around the time that army ants originated in Africa, a separate lineage of dorylines colonized the New World. This lineage turned out to be incredibly successful, giving rise to the vast majority of modern New World dorylines, including all New World army ants. A second, ecologically much less impactful colonization of the Americas by dorylines occurred substantially later and resulted in the New World species of the inconspicuous genus *Syscia* (Borowiec 2016, 2019) (Figure 2.15).

The Old World army ants are a group encompassing the genera *Aenictus* (with currently 187 valid species according to AntWeb.org), *Aenictogiton* (seven species), and *Dorylus* (sixty species). *Aenictus* is the most species-rich genus of army ants, and prolific recent taxonomic efforts suggest that many species remain to be discovered (e.g., Shattuck 2008; Liu et al. 2015; Jaitrong and Ruangsittichai 2018). The bulk of the biodiversity in the genus *Aenictus* is found in tropical and subtropical Asia, but several species occur in Africa (e.g., Wilson 1964) (Figure 2.16). Ranging in the lower tens of thousands, their colony sizes are fairly small for army ants. *Aenictus* workers are usually monomorphic, and most species feed exclusively on other ants. However, at least one species, *Aenictus inflatus,* has a polymorphic worker caste. The name refers to a curious inflation on the back of large workers, filled with a red liquid of unknown function (Yamane and Hashimoto 1999). Other species, such as *Aenictus laeviceps,* are rather conspicuous, with larger colonies, surface raiding, and a somewhat broader prey spectrum (see Table 2.3).

The genus *Aenictogiton* is arguably the most enigmatic among the army ants. Long only known from males, the first workers have only recently been described, and virtually nothing is known about the ecology and behavior of these ants other than that they appear to be strictly subterranean predators (Borowiec 2016). In fact, the designation of *Aenictogiton* as army ants currently rests solely on their phylogenetic proximity to *Aenictus* and *Dorylus* (see Figure 2.13). *Aenictogiton* is only known from Africa.

The biogeographic center of the genus *Dorylus* is likewise in tropical Africa, although some species also occur in Asia, including *Dorylus laevigatus,* the sister species to all other *Dorylus* (Brady 2003; Kronauer et al. 2007c). With several hundreds of thousands to several million workers, the colonies of *Dorylus* are enormous, and the worker caste of many species shows a striking level of polymorphism (see Table 2.3). Unlike *Aenictus,* species in the genus *Dorylus* are usually generalist predators rather than specialized on ant prey, even though they do include social insects in their diet, and some feed predominantly on termites (Gotwald 1978, 1995) (Figure 2.17; see Table 2.3). The most recent common ancestor of all *Dorylus* army ants was arguably strictly subterranean and most species still are, but the African driver ants have evolved the largest colony sizes among army ants and conduct massive surface swarm raids (Kronauer et al. 2007c) (Figure 2.18).

FIGURE 2.16 *Aenictus* is the most diverse army ant genus, with most species occurring in tropical Asia. However, a few are also known from Africa. This picture shows *Aenictus* workers (probably of the species *Aenictus eugenii*, a predator of myrmicine ants; Gotwald and Cunningham-van Someren 1976) standing guard at the periphery of a colony emigration. In the background, workers carrying larvae emerge from a hole in the ground, only to disappear again a couple of centimeters later. This colony was encountered right after dusk at Mpala Research Centre in Kenya. While emigrating, the ants also conducted a conspicuous surface raid.

FIGURE 2.17 Most army ants lead a subterranean existence and are rarely encountered above ground. These *Dorylus fimbriatus* ants briefly came to the surface while attacking a beetle grub. Army ants in the Old World genus *Dorylus* and the New World genus *Labidus* often prey on a broad range of invertebrates, while other doryline army ants and most non–army ant dorylines primarily attack the colonies of other ants. Kakamega Forest, Kenya.

FIGURE 2.18 African driver ants are highly polymorphic and readily observed on the surface. This image shows the variously sized workers of *Dorylus molestus* form a column across a foot path in the tea plantations at Mount Kenya, Kenya.

The New World army ants contain the genera *Cheliomyrmex* (four species), *Eciton* (twelve species), *Labidus* (seven species), *Neivamyrmex* (126 species), and *Nomamyrmex* (two species). The oldest known army ant fossils, representatives of the genus *Neivamyrmex,* belong to this clade (Wilson 1985; Barden 2017) (Figure 2.19). *Neivamyrmex* fossils have been found in Dominican and Chiapas amber deposits, both on the order of 19 million years old, providing a minimum age estimate for New World army ants and the genus *Neivamyrmex* in particular (Wilson 1985; Coty et al. 2014; Borowiec 2019). These amber pieces are direct testimony to the ancient ways of the army ants: one contains a *Neivamyrmex* worker holding a prey termite, and another one entrapped a worker carrying a wasp pupa (Poinar and Poinar 1999; Coty et al. 2014).

Although army ants are absent from the Dominican Republic and the Greater Antilles in general today, the genus *Neivamyrmex* is by far the most speciose among the New World army ants and remains widely distributed. Several species are found as far north as the southern United States, and new species are still described occasionally (e.g., Watkins 1985; Snelling and Snelling 2007; Varela-Hernández and Castaño-Meneses 2011). Relative to other army ants, *Neivamyrmex* have small to medium-sized colonies and low levels of worker polymorphism (Figure 2.20; see Table 2.3). The genus *Cheliomyrmex* is the most subterranean, and thus most elusive among the New World army ants. Workers are rarely encountered, and queens are still unknown. Species in the genera *Labidus, Nomamyrmex,* and *Eciton* can be more conspicuous and are distributed throughout the tropical and subtropical regions of the Americas. Several species in these genera have very large colonies and raid at least partly above ground, and species in the genera *Labidus* and *Eciton* also have highly polymorphic workers that can differ dramatically in overall size and appearance (see Table 2.3). This increased level of morphological specialization observed in army ants with particularly large colonies is thought to be facilitated by the fact that large societies rely less on the behavioral flexibility of individual workers (Bourke 1999).

FIGURE 2.19 The only known army ant fossils belong to the genus *Neivamyrmex*, such as this inclusion in Dominican amber, ca. 19 million years old.

FIGURE 2.20 Doryline army ants in the two most species-rich genera, *Neivamyrmex* and *Aenictus*, typically have low levels of worker polymorphism and are specialized ant predators. This image shows *Neivamyrmex pilosus* workers returning from a raid carrying pilfered ant brood.

In keeping with doryline tradition, most New World army ants are primarily myrmecophagous, and some show an astonishing level of specialization on one or a few prey species (e.g. Mirenda et al. 1980; Rettenmeyer et al. 1983; Powell and Baker 2008; Hoenle et al. 2019). However, a few have evolved broader prey spectra, and especially species in the genus *Labidus* are reminiscent of their distant *Dorylus* relatives in terms of their catholic food choices (Figure 2.21).

FIGURE 2.21 Among the New World army ants, species in the genus *Labidus* arguably have the broadest prey spectra. This image shows a large *Labidus coecus* worker.

Building on what we know about the phylogenetic relationships between the different groups of ants, as well as the taxonomic distribution of army ant biology, we will now look at how army ant behavior has evolved in different lineages. Although army ant–like behavior is most common in the Dorylinae, it also occurs in several species of ponerines, where it has likewise been studied in substantial detail (see Table 2.1). As we will see, doryline and ponerine army ants have evolved along surprisingly similar trajectories in a striking case of evolutionary convergence.

The Evolution of Spontaneous Mass Raiding

Many ants are predatory. Some kill and eat other animals as part of a broader diet, and others refuse to take anything but live prey. But the foraging strategies of ants are highly diverse, and not all predatory ants classify as army ants (Dornhaus and Powell 2010; Cerdá and Dejean 2011; Lanan 2014). The weaver ants of tropical Africa and Asia, for example, form large colonies consisting of many nests that the ants "weave" out of leaves, using their larvae as silk-producing shuttles (see Figure 2.6). Weaver ant colonies can span the canopies of several large trees, and the different nests are connected by a network of trails that the ants defend viciously against intruders. While weaver ants rely heavily on the sugary excretions of herbivorous insects for their diet, they will also attack, kill, and consume most arthropods that wander into their territory (Hölldobler and Wilson 1990; Blüthgen and Fiedler 2002) (see Figure 2.7). Once an intruder has been detected, more ants are recruited from the vicinity via volatile pheromones to assist in the attack (Hölldobler 1983; Hölldobler and Wilson 1990). Many species of the ponerine genus *Leptogenys,* on the other hand, form small colonies, are highly specialized in their prey spectrum, which often consists exclusively of isopods, and practice solitary foraging. Foragers sting isopods upon encounter and transport them back to the nest (Maschwitz and Steghaus-Kovac 1991; Freitas 1995; Dejean and Evraerts 1997; Hoenle 2019).

What sets army ants apart from these and other predatory ants is that their colonies do not defend stable territories or send out solitary foragers.

Instead, army ants leave the nest en masse and in a coordinated manner to go hunting, and they find, overwhelm, and kill their prey in large groups (Schneirla 1971; Gotwald 1995; Powell and Baker 2008). Importantly, the raiding party leaves the nest without any knowledge of where prey is located, and the ants encounter prey as the raid advances. An army ant raid therefore has no distinct search phase, and recruitment upon prey encounter happens at the raid front. At least among the dorylines, this form of foraging, which I call "spontaneous mass raiding," is restricted to species with colonies that have tens of thousands of workers or more. (The terminology describing ant foraging behavior has not always been used consistently, and I am loosely following Lanan 2014; see Reeves and Moreau 2019 for a recent overview.) In tropical Asia, spontaneous army ant mass raiding also occurs in some ponerines of the genus *Leptogenys* as well as the marauder ants in the subfamily Myrmicinae. However, even some ponerines with smaller colonies conduct spontaneous raids (see Table 2.1). In fact, in the ponerines, and to a lesser extent in the dorylines as well, species fall along a continuum of army ant–like behavior, and the definition of what exactly constitutes an army ant therefore becomes a bit arbitrary. Although this gradation thus does not always allow us to pigeonhole nature, it beautifully illustrates the likely evolutionary trajectory from solitary foraging to mass predation.

This is particularly true for the various stages of army ant–like behavior observed in the ponerines (Wilson 1958a; Hölldobler and Wilson 1990; Maschwitz and Steghaus-Kovac 1991; Kronauer 2009, 2020). Even in some of the *Leptogenys* species that usually send out lone foragers, recruitment of nestmates can occur when an ant happens upon a cluster of several isopods or a large prey item that she cannot retrieve by herself (Maschwitz and Steghaus-Kovac 1991; Dejean and Evraerts 1997). This behavior becomes the norm in species that routinely attack prey that cannot be overwhelmed by a single ant. These can be large arthropods or in some cases the colonies of other social insects, particularly termites. During recruitment, the scout that has successfully discovered a target lays a pheromone trail on her way back to the nest and alerts other workers upon her return (Lanan 2014). This elicits what I call a "scout-initiated group raid," which can involve a large proportion of the colony's members. In contrast to a spontaneous mass raid, a scout-initiated group raid has a distinct search phase, during which one or more scouts explore outside the nest. The actual raiding party, on the other hand, only leaves the nest once prey has been discovered, and it travels directly in the direction indicated by the scout.

Species with scout-initiated group raiding typically live in larger colonies than related species with solitary foraging. In *Leptogenys diminuta,* for example, colonies measure on the order of 300 workers, and scouts recruit raiding parties anywhere from three to over a hundred strong to seize cockroaches, mantids, or large spiders (Maschwitz and Mühlenberg 1975; Attygalle et al. 1988; Maschwitz and Steghaus-Kovac 1991; Ito and Ohkawara 2000; Witte et al. 2010) (see Table 2.1; see Figure 2.2). However, in terms of size, the colonies of species with scout-initiated group raiding never approach those of the species we consider army ants. The terms "group raid" and "mass raid" thus simply denote raids of different sizes, typically not more than a thousand workers in group raids, and tens of thousands of workers or more in mass raids. As a corollary of the large number of workers involved, army ant mass raids usually remain connected to the nest by a column of workers that travel back and forth while the raid is in progress, whereas group raids break away from the nest.

Among the ponerines, scout-initiated group raiding has been particularly well-studied in *Megaponera analis,* a specialist predator of fungus-growing termites of the subfamily Macrotermitinae (Longhurst and Howse 1979; Longhurst et al. 1979a; Villet 1990; Hölldobler and Wilson 1990; Hölldobler et al. 1994b; Corbara and Dejean 2000; Bayliss and Fieldling 2002; Yusuf et al. 2014a, 2014b; Frank and Linsenmair 2017a, 2017b; Frank 2020) (see Table 2.1). Colonies of this species measure anywhere from several hundred to over a thousand workers in strength. Unlike army ants, *Megaponera analis* colonies initially send out scouts to scope out termite foraging galleries. Once a scout has discovered foraging termites, she rushes back to her colony along the fastest route while depositing a pheromone trail (Frank et al. 2018a). A minute or so after entering her nest she emerges again, now leading a raiding column of a few dozen to a few hundred workers back to the food source. The role of the scout in this process is paramount: if she is removed, the raiding party becomes disoriented, disbands, and the ants return home empty-handed (e.g., Longhurst and Howse 1979; Hölldobler et al. 1994a; Bayliss and Fielding 2002). These raids can cover distances of up to 40 meters.

Once the party has reached its destination, the ants gather in a cluster before the largest workers begin to break open the earthen termite galleries. The smaller workers then rush into the galleries, capture termites, paralyze them by stinging, and deposit them in piles at the entrance to the gallery. Now their larger sisters take over once again, picking up the bounty in bundles of up to seven termites at a time to carry them back to the nest (Figures 2.22 and 2.23). These battles with well-defended insect societies are not without cost, and ants often get injured while fighting with termites. However, the army comes with its very own emergency and hospital service. A worker that has lost one or a few legs in combat emits a pheromone that triggers rescue behavior. Not only is the injured combatant carried back to the nest by her nestmates, but her wounds are cleaned once the raiding party has returned home, which dramatically decreases the risk of infection and increases survival. Fatally wounded ants, on the other hand, do not trigger the rescue behavior and die on the battlefield (Frank et al. 2017, 2018b). *Megaponera analis* workers also have each other's back during the occasional encounter with *Dorylus nigricans* driver ants, where they remove army ant workers that cling to the appendages of their nestmates (Beck and Kunz 2007).

Scout-initiated group raiding is probably also predominant in non–army ant dorylines. The biology of representative species with somewhat detailed records is summarized in Table 2.2, and patchier observations of additional species can be found elsewhere (e.g., Wheeler 1918; Clark 1923, 1924, 1941; Wilson 1958b, 1959; Brown 1975; Briese 1984; Mackay 1996; Brandão et al. 1999; Mariano et al. 2004; Donoso et al. 2006; Bolton and Fisher 2012; Fisher and Peeters 2019; reviewed in Borowiec 2016). Unfortunately, given how rarely most species are encountered, there are currently only three species for which entire raids have been documented: *Cylindromyrmex whymperi* (Gobin et al.

FIGURE 2.22 During the attacks of *Megaponera analis* on termite galleries, small and large workers efficiently divide tasks. The small workers enter the galleries, retrieve termites one at a time (on the right in the background), and pile them up on the ground (on the left in the background). The large workers then pick up the termites in bundles (foreground) and carry them back to the nest. Mpala Research Centre, Kenya.

FIGURE 2.23 On their homebound journey from a raid, *Megaponera analis* workers form a column in which large workers carry bundles of termites back to the nest. Many of the small workers do not carry anything (bottom center). Mpala Research Centre, Kenya.

2001), *Lioponera* cf. *turneri* (Hölldobler 1982), and the clonal raider ant *Ooceraea biroi* (Chandra et al. n.d.) (see Table 2.2).

In *Lioponera* cf. *turneri,* individual scouts leave the nest to explore the surroundings for prey ant colonies, particularly those of the genus *Pheidole* (Hölldobler 1982). Once a scout has encountered a *Pheidole* colony, she immediately returns home, marking her route with secretions from the poison gland that serve as a trail pheromone. Upon entering her nest, she releases a second, volatile recruitment pheromone from a different gland. The communicated excitement rapidly spreads among her nestmates, and a raiding party now exits the nest and picks up the pheromone trail that leads to the unsuspecting prey. What follows is breaking and entering. Even though they are vastly outnumbered, the heavily sclerotized raiders overwhelm the *Pheidole* workers and soldiers without much effort, stinging and paralyzing them. Immobilized adults and their brood are then carried back to the *Lioponera* nest, with workers running back and forth until the *Pheidole* colony is thoroughly plundered (Hölldobler 1982).

Because the clonal raider ant can be maintained in the laboratory in large numbers and for extended periods of time, this species is particularly suitable to study the intricacies of doryline behavior (Figure 2.24). The raids, which resemble those of *Lioponera* cf. *turneri* in overall structure, can reliably be elicited to isolated prey items, such as pupae of the fire ant *Solenopsis invicta* (Figure 2.25). Interestingly, unlike in *Megaponera analis* and some other group-raiding ponerines (e.g., Bayliss and Fielding 2002; Mill 1984), the scout usually does not lead the raiding party, and is thus not required for the ants to orient themselves toward the prey (Chandra et al. n.d.). The scout's pheromone trail suffices. This decreased dependency on the scout during the raid is thought to be more efficient because it can translate into more continuous and persistent recruitment (Chadab and Rettenmeyer 1975). The type of recruitment practiced by the clonal raider ant and possibly other non–army ant dorylines might therefore facilitate the evolution of army ant mass raiding (Chandra et al. n.d.). Indeed, even though army ant raids are not elicited by scouts, recruitment events reminiscent of clonal raider ant group raids can still be observed at army ant raid fronts, as we will see in Chapter 3 (Chadab and Rettenmeyer 1975).

FIGURE 2.24 The clonal raider ant, *Ooceraea biroi*, is among the few dorylines that can be readily maintained in the laboratory for experimentation. The workers shown here with larvae are tagged with unique combinations of color dots. This allows us to follow individuals over time and monitor their behavior. Laboratory colony collected on St. Croix, U.S. Virgin Islands, and maintained at the Rockefeller University.

This opens the question of how scout-initiated group raiding transitioned to spontaneous mass raiding as colony size increased over evolutionary time. One approach to address this problem is to study the raiding behavior of species with intermediate colony sizes. For the longest time, however, the jump in colony size from scout-initiated group raiding to spontaneous mass raiding species appeared so dramatic and discrete in the dorylines that army ants truly seemed to be in a league of their own (see Tables 2.2 and 2.3). Yet, at least in principle, colony size is a continuous trait. This has become apparent with the recent discovery of the doryline genus *Yunodorylus* in tropical Asia (Xu 2000; Borowiec 2009). The colonies of *Yunodorylus eguchii,* for example, contain several thousand workers, exactly filling this gap (Eguchi et al. 2016; Riou Mizuno and Fuminori Ito, personal communication) (see Table 2.2). Even though observations on the raiding behavior of *Yunodorylus eguchii* are still pending, the first colonies have successfully been maintained in the laboratory, opening an exciting opportunity to finally gain

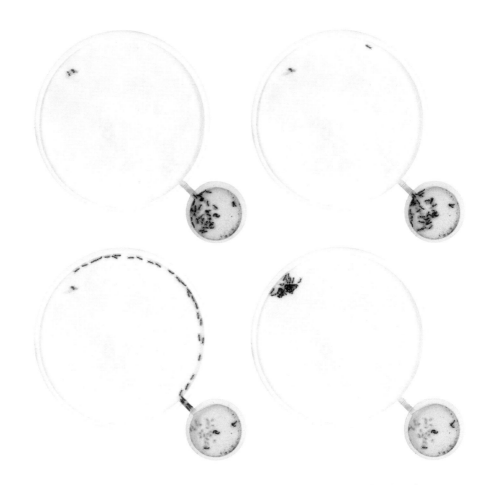

FIGURE 2.25 Different stages of a scout-initiated group raid in the clonal raider ant. *Top left:* A solitary scout has encountered prey (pupae of the fire ant *Solenopsis invicta* in this case). *Top right:* The scout returns to the nest laying a pheromone trail, but without actually retrieving prey. Upon entering the nest, she will release a volatile recruitment pheromone. *Bottom left:* In response to the scout's recruitment phero-mone, a large proportion of the workers in the colony exit the nest and follow the pheromone trail. Note that the scout, who can be identified by the color tag, is not leading the raiding party. *Bottom right:* The raiding party has reached its destination, and will soon begin retrieving the prey. In nature, recruitment would most likely have occurred to an actual prey ant colony. Laboratory colony collected on St. Croix, U.S. Virgin Islands, and maintained at the Rockefeller University.

much needed insight into this crucial step in army ant evolution (Eguchi et al. 2016).

Another approach is to experimentally increase colony size in species with scout-initiated group raiding and to study the effect on raid onset. Pursuing this avenue in the clonal raider ant, we found that even though the vast majority of raids are initiated by scouts, colonies also show spontaneous bouts of activity in which groups of workers leave the nest and, in the absence of trail pheromone, hesitantly venture farther and farther into the foraging arena. This behavior strikingly resembles what Theodore Schneirla called "pushing parties": small groups of workers that venture a few centimeters from the bivouac at the onset of an army ant column raid, depositing trail pheromone before retreating again (Schneirla 1933; Schneirla and Reyes 1966). The members of a pushing party are soon replaced by new ants, which now more readily cross the previously traveled terrain, only to hesitantly extend the explored territory by another few centimeters before turning back (Schneirla 1933). Importantly, in our experiments, these synchronized bouts of spontaneous activity were more frequent in larger colonies. Our largest experimental colonies, which were an order of magnitude larger than naturally occurring colonies of this species, even conducted spontaneous mass raids: recruitment occurred at the raid front, which was connected to the nest by a constant stream of ants (Chandra et al. n.d.). This shows that spontaneous raids can emerge from excitatory interactions between colony members even in species that usually conduct scout-initiated raids, and that these events increase in frequency as a function of colony size. Furthermore, it suggests that the behavioral rules followed by individual ants have not changed much in army ants, and that army ant societies, rather than constituting a clearly distinct category, are essentially just massively scaled-up versions of those of their more humble relatives (Chandra et al. n.d.).

As we have seen above, army ants with spontaneous mass raids have probably evolved twice in the dorylines, and each of the two clades has subsequently diversified into more than a hundred species. In contrast, army ant raiding behavior in the ponerines has only been described for a few conspicuous Asian species in the genus *Leptogenys,* all of which have broad food spectra (Maschwitz and Steghaus-Kovac 1991). The best-studied ponerine army ant is *Leptogenys distinguenda* (Maschwitz et al. 1989; Maschwitz and Steghaus-Kovac 1991; Witte and Maschwitz 2000, 2002; Hoenle 2019) (see Table 2.1). Recapitulating what we have seen in the dorylines, colonies of this species are much larger than those of closely related species with scout-initiated group raiding, measuring on the order of 50,000 workers. Even though this number is at the low end of doryline army ant colony sizes, it is well within the range. Colonies construct temporary bivouacs in preexisting subterranean cavities or inside hollow tree trunks and can form true swarm raids. With about 10 meters in width and 15 meters in depth, these swarms resemble those of *Eciton burchellii* in terms of ontogeny, organization, and dimension (Maschwitz et al. 1989; Maschwitz and Steghaus-Kovac 1991). The species is not picky when it comes to prey, overwhelming and killing earthworms, snails, a variety of large arthropods including scolopenders, and even the occasional frog or snake (Maschwitz et al. 1989).

The evolution of scout-initiated group raiding, the precursor of spontaneous army ant mass raiding, has frequently gone hand in hand with the evolution of two additional traits: nomadism and dependent colony founding by fission with wingless queens (see Tables 2.1–2.3). Collectively, mass raiding, nomadism, and colony fission constitute what is known as the "army ant adaptive syndrome" (see also Wilson 1958a; Schneirla 1971; Gotwald 1978, 1988, 1995; Hölldobler and Wilson 1990; Brady 2003; Kronauer 2009, 2020). The following three chapters will discuss each trait separately and in detail, but it is useful to begin with a brief overview of how they are functionally and evolutionarily interconnected.

The Evolution of Nomadism and Colony Fission

Army ants and their relatives rely on ephemeral and unpredictable food sources that can become locally depleted, such as social insect colonies or large solitary arthropods. The problem of running out of food applies especially to species that are dietary specialists and / or have large colonies that demand large quantities of prey. Early in army ant evolution, this selective pressure arguably gave rise to a trajectory away from the sessile habits of most ants toward a nomadic lifestyle. Accordingly, nomadism is the norm among group raiding ponerines (see Table 2.1), and it has even been reported for *Leptogenys* species with solitary foragers (Maschwitz and Steghaus-Kovac 1991; Freitas 1995). This suggests that nomadism can in fact predate group raiding along the evolutionary trajectory toward becoming an army ant, and it has been suggested that, at least in *Leptogenys,* group recruitment during colony relocation served as a preadaptation to group raiding (Maschwitz and Steghaus-Kovac 1991).

Nomadism is also common, if not ubiquitous, among non–army ant dorylines (see Table 2.2). Direct observations of colony emigrations in non–army ant dorylines have only been made in a few cases, but the nests of many species appear to be impermanent, suggesting that colonies relocate readily and frequently (Brown 1975). In a subset of doryline army ants and their relatives, colonies undergo stereotypical cycles, alternating between "reproductive" or "statary" phases and "brood care" or "nomadic" phases. In these phasic species, colony emigrations occur only during the nomadic phase (see Tables 2.2 and 2.3). Outside of the dorylines, phasic colony cycles are either absent or have not been well documented (see Table 2.1). These cycles and their putative adaptive value will be discussed in detail in Chapter 4.

The idea that transient food sources are evolutionarily linked to the evolution of nomadism is supported by a few additional instances in which ants have become nomadic. The formicine *Euprenolepis procera,* for example, harvests mushrooms in the rainforests of Southeast Asia. The mushroom fruiting bodies are short-lived, and their appearance is erratic. Consequently, the ants are nomadic, and colony emigrations in fact seem to be directly triggered by the local depletion of mushrooms (Witte and Maschwitz 2008; von Beeren et al. 2014). The lifestyle of the dolichoderine *Dolichoderus cuspidatus* is similarly unconventional. This "herdsmen" species tends phloem sap-feeding mealybugs, which the ants frequently relocate to sparsely distributed sprouting plants. Colonies measure over 10,000 workers and construct temporary bivouac nests not unlike those of army ants. Once the distance between the bivouac and the feeding site has become too large, the ants gather their livestock and brood to move and set up camp elsewhere (Maschwitz and Hänel 1985; Dill et al. 2002).

However, these herdsmen face another challenge: their livelihood relies entirely on honeydew, the sugary excretion harvested from their mealybugs, and colonies without mealybugs are inviable. So how do you make sure that incipient colonies have mealybugs from the get-go? In some ant species that rely heavily on honeydew, the young queens bring along a starting stock of cattle on their mating flight before founding a new colony independently. The herdsmen, however, have found another ingenious solution: when colonies have reached a certain size, they simply split into two, with each retaining a decent share of mealybugs, along with a single, permanently wingless queen (Maschwitz and Hänel 1985; Dill et al. 2002). In this mode of colony reproduction, called dependent colony founding, the queens rely on the assistance of workers to establish new colonies.

Dependent colony founding has evolved repeatedly across the ant phylogeny under conditions where secluded queens or tiny incipient colonies are not viable (Cronin et al. 2013). As group predation became obligate in the ancestors of army ants, independent colony founding was no longer an option—an army had to be of a certain size in order to succeed (Wilson 1958a; Schneirla 1971; Franks and Hölldobler 1987; Gotwald 1988, 1995; Hölldobler and Wilson 1990; Brady 2003; Kronauer 2009, 2020). But because ant workers cannot fly, the young queens were now likewise bound to the ground and had to disperse on foot. This change in lifestyle allowed army ant queens to undergo additional modifications that set them apart from the queens of most other ants.

First, army ant queens are born without wings, and in this sense they are "ergatoid," meaning "worker-like" in Greek. Second, in species with very large colonies, including all doryline army ants, the queens have massively expanded ovaries and gasters, along with a few other derived morphological traits (Borowiec 2016; Hölldobler 2016). For example, while the queens of many non–army ant dorylines have small ovaries with only two ovarioles each, each ovary of an *Eciton burchellii* queen contains approximately 1,300 such tubes (Hagan 1954a, 1954b, 1954c; Ito et al. 2018). This exaggeration of queen traits might well be incompatible with the ability to fly, and thus only became possible once dependent colony founding had evolved. Army ant queens have truly become über-mothers, or "dichthadiigynes," from the Greek words *dichthadios,* meaning double, and *gyne,* meaning woman (Wheeler 1908).

Among the non–army ant dorylines, a large spectrum of queen phenotypes can be observed, ranging from winged queens, to wingless queens that resemble the workers in size and shape, to wingless queens that begin to approach true dichthadiigynes (Borowiec 2016) (see Figure 2.15; see Table 2.2). *Yunodorylus* queens, for example, are classified as "subdichthadiiform" (Eguchi et al. 2016; Satria et al. 2018). In other words, they represent a light version of the exaggerated doryline army ant queens. The queens of several non-doryline species with army ant characteristics in the genera *Leptanilla, Onychomyrmex,* and *Simopelta* also have ergatoid queens with appreciable levels of dichthadiigyny (Kronauer 2009; Hölldobler 2016) (see Table 2.1). When exactly ergatoid queens and dependent colony founding evolved along the evolutionary trajectory toward army ants is again difficult to say. Most *Leptogenys* species with solitary foragers, for example, already have ergatoid queens, while among the non–army ant dorylines, even some group-raiding species still possess winged queens (Lattke 2011; Rakotonirina and Fisher 2014; Schmidt and Shattuck 2014; Borowiec 2016) (see Table 2.2).

As we have seen in this chapter, species with army ant–like traits have evolved repeatedly across the ant phylogeny. However, species that display all the traits of the army ant adaptive syndrome—spontaneous mass raiding, nomadism, and reproduction by colony fission—are restricted to the ponerine genus *Leptogenys* and two clades in the subfamily Dorylinae, the latter of which contain the vast majority of extant army ant species. Interestingly, in each instance, the evolution of army ants was a gradual process that unfolded along strikingly similar trajectories from species with small to medium-sized colonies that conducted scout-initiated group raids, to species that form among the largest insect societies on Earth and hunt in spontaneous mass raids (see also Wilson 1958a; Gotwald 1985, 1988, 1995; Hölldobler and Wilson 1990; Kronauer 2009, 2020).

CHAPTER THREE

Mass Raiding

Woken by the ear-shattering guttural growls of howler monkeys, my field companion Christoph von Beeren and I have decided to visit a colony of the swarm raider *Eciton burchellii* that is bivouacked between the buttress roots of a tall tree at La Selva Biological Station. Around dawn the air is still cool and refreshing, and the forest is drenched from last night's heavy rains. The surface of the bivouac is calm, and only a few individual ants are rummaging around on the colony's garbage heap. We sit down on our folding stools and wait, watching birds hopping around in the vicinity. The first rays of sunlight, however, excite the ants, and soon an ever increasing number are milling about around the base of the tree. At around 7:00 a.m., a shiver of activity spreads through the bivouac. The beast is about to rise, and a new swarm raid seems imminent. Sure enough, only a few seconds later, a giant wave of ants spills out from the bottom of the bivouac and onto the forest floor, where the masses spread out into a dense carpet. Within a few minutes, the swarm assumes a more discrete direction and begins moving away from the nest. We hastily fold up our stools and get out of harm's way.

Shortly before 8:00 a.m. the raiding party enters the clearing of the research station, and a few minutes later, it reaches Cabina Eufonia, the bungalow closest to the forest edge. The ants storm the concrete veranda at ground level and overrun a large carpenter ant colony whose nest is located somewhere in the foundation of the building. The docile and nocturnal carpenter ants offer little opposition to the onslaught, and the veranda is soon littered with casualties (Figure 3.1). While the fight on the veranda continues, the army ascends the outer walls of the bungalow

(Figure 3.2). The bungalows at La Selva Biological Station are equipped with exterior lighting that attracts insects at night, and some of them usually still hang out around the lightbulbs in the morning. These unfortunate stragglers are next in the line of attack, and a few rudely awakened katydids jump from the ceiling overhang straight into the pandemonium raging below, where they are immediately pinned down by dozens of army ants, a scene reminiscent of *Gulliver's Travels* (Figures 3.3).

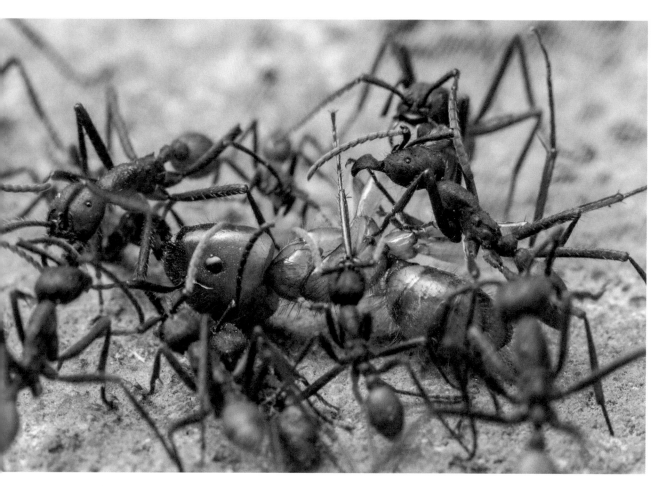

FIGURE 3.1 Most army ants are specialist predators of other ants. Here, *Eciton burchellii* workers are attacking a large *Camponotus atriceps* (subfamily Formicinae) worker on the concrete porch of a bungalow at La Selva Biological Station, Costa Rica.

The human inhabitants of the bungalow have left the station for a short trip, and we are thus unfortunately not able to witness the fighting as it continues inside. However, when I stop by again after dinner that day, nearly twelve hours later, dead carpenter ants are still being carried down along the column that had ascended the bungalow in the morning, and the resident researchers will find their beds littered with ant corpses upon their return the next day.

FIGURE 3.2 Early naturalist accounts frequently include observations of *Eciton burchellii* entering human dwellings to hunt. Here, a raid is storming Cabina Eufonia at La Selva Biological Station, Costa Rica.

The battle described here was clearly an epic one, though far from unprecedented, and reminiscent of Maria Sibylla Merian's description of *Eciton burchellii* raids entering human dwellings. Their welcome role as a cleaning force has been part of the narrative ever since (Figure 3.4).

FIGURE 3.3 During a raid on a bungalow at La Selva Biological Station in Costa Rica, *Eciton burchellii* army ants are attacking a katydid nymph (*Clepsydronotus* spec.; subfamily Pseudophyllinae).

THE SPRING CLEANING.

Vast hordes of foraging ants visit a house in Trinidad and drive out or kill all the vermin—cockroaches, bugs, West Indian species, mice, rats, and bug wasps—all the fleeing ones having numerous ants attached to them by the jaws, so that they were reminded to the painful bites.

137

FIGURE 3.4 Similar to Merian's account of *Eciton burchellii* entering houses and killing insects and spiders, the army ants appear in a rather positive light in the account of a Mrs. Carmichael from Trinidad: "Ah, misses, you've got the blessing of God today, and a great blessing it is to get such a cleaning" comments one of the spectators as the ants finally leave the house of Mrs. Carmichael carrying off their booty. The full account is reproduced in the classic *An Introduction to Entomology* by Kirby and Spence (1843) and is attributed to Mrs. Carmichael ("On the West Indies," quoted in *Saturday Magazine*, 1833, p. 150). The account was later partially included in Edward Step's 1916 book *Marvels of Insect Life*, and the accompanying illustration "The Spring Cleaning" (137) is reproduced here. Note that this depiction is inaccurate in some biological details. For example, large arthropods like cockroaches and crickets are not transported in one piece as is shown here, but chopped into smaller parts on the spot. Furthermore, *Eciton burchellii* hardly ever preys on vertebrates, and the mouse being carried by the ants in one piece is clearly a product of the artist's imagination. Image from the Smithsonian Libraries / Biodiversity Heritage Library.

Raid Organization

Large carpet swarms are only employed by generalist predators whose diet includes a broad range of solitary arthropods. Among the dorylines, these include *Eciton burchellii* and *Labidus praedator* in the Neotropics, as well as *Dorylus* driver ants in Africa. In Asia, the ponerine *Leptogenys distinguenda* and the marauder ants in the myrmicine genus *Carebara* also hunt in large swarms. By contrast, the vast majority of army ants, whose diet is limited to social insects, usually raid in columns. As we have discussed in Chapter 2, this difference in diet also correlates with differences in colony size, in that swarm raiders with broader diets generally also have larger colonies. And just like army ant colony sizes, raiding patterns fall along a continuum (Rettenmeyer 1963a; Burton and Franks 1985; Gotwald 1995). I will begin this section with a discussion of the massive swarm raids of *Eciton burchellii,* along with some of the general collective phenomena displayed by army ants. I will then treat *Eciton hamatum* as a representative column raider and conclude by arguing that, just as the transition from scout-initiated group raiding to spontaneous mass raiding, the transition from army ant column to swarm raiding constitutes an emergent phenomenon contingent upon gradual increases in colony size.

A fully developed *Eciton burchellii* swarm raid measures on average around 12 meters across, and large swarms can measure 20 meters (Willson et al. 2011 for data from the Peruvian Amazon). The swarm is around 2 meters deep and composed of many thousand little warriors (Figure 3.5). The advance of swarm raids, both in *Eciton burchellii* and African *Dorylus* driver ants, is reminiscent of the pushing parties in column raids discussed in Chapter 2: ants that reach the swarm front venture hesitantly into uncharted territory, extend the film of surface pheromone by a few centimeters, and then turn back (Schneirla 1940; Leroux 1977a, 1982). In fact, it is this deceleration of the stream of outbound ants reaching the swarm front that causes a back pressure and the spreading out of the swarm (Schneirla 1940). It is thus the magnitude of the outbound flux of excited ants from the bivouac that determines both the dimension and the impetus of the carpet swarm.

FIGURE 3.5 The ants at the front of an *Eciton burchellii* swarm raid are highly aggressive and ready to attack, as is evidenced by their wide open mandibles. An *Eciton burchellii* swarm in its natural rainforest environment is essentially impossible to photograph in its entirety, and this picture therefore only shows a tiny section of the swarm.

The running speed of *Eciton* army ants along established trails depends on worker size and temperature, and is somewhere between 5 and 13 centimeters per second (Bartholomew et al. 1988; Powell and Franks 2005; Meisel 2006; Hurlbert et al. 2008). Yet given the hesitation of ants at the swarm front, even at a maximum speed of approximately 12 to 14 meters per hour, the swarm advances only slowly (Schneirla 1940). Throughout the course of the day, however, it can move over 100 meters away from the bivouac, hauling in more than 30,000 prey items (Rettenmeyer 1963a; Schneirla 1971; Rettenmeyer et al. 1983; Franks 1985; Franks et al. 1991; Vieira and Höfer 1994; Willson et al. 2011).

When an *Eciton* worker encounters prey at the raid front, she elicits a strong recruitment response from her nearby sisters, probably via a volatile pheromone that acts at short distance (Chadab and Rettenmeyer 1975; Topoff and Mirenda 1975). Consequently, it only takes a few seconds until large prey arthropods at an *Eciton burchellii* swarm front are covered in ants (Figure 3.6). Upon closer inspection, however, it becomes apparent that, despite the massive amplification in numbers, the descent of army ants from ancestors with scout-initiated group raiding also shines through at the raid front. When an *Eciton* worker detects prey, such as a social insect colony, at some distance from the main raid, she returns to the raid depositing a pheromone trail. Once she reaches the raid, she runs around among her nestmates and alerts them to the recruitment trail, possibly once again via a volatile recruitment pheromone that acts at short distance. She might also conduct additional trips along the path leading to the prey, seemingly reinforcing the pheromone trace. Ants that have been in contact with this primary recruiter can act as secondary recruiters by activating additional ants. Within a minute of the recruiter's arrival at the raid, the first cohort of fifty to 100 army ants arrive at the prey, ready to attack (Chadab and Rettenmeyer 1975).

The recruitment trail laid by a single scout in combination with active recruitment upon the scout's arrival is capable of diverting *Eciton* workers from heavily reinforced foraging trails, suggesting that the chemical composition of recruitment trails might in fact differ from that of regular foraging trails (Chadab and Rettenmeyer 1975; Topoff et al. 1980a;

FIGURE 3.6 Army ants, like these *Eciton burchellii*, overwhelm their victims en masse, and within seconds even large insects like this katydid nymph (*Idiarthron hamuliferum*; subfamily Pseudophyllinae) are entirely covered by the ants.

Topoff 1984; Billen and Gobin 1996). This behavioral sequence, even though it plays out at the raid front rather than the nest, is quite reminiscent of the group raids conducted by non–army ant dorylines, as discussed in Chapter 2. However, a pheromone trail is not strictly necessary for local recruitment from the raid front. When a scouting *Eciton* worker discovers a suitable social insect prey colony up in the vegetation, she may simply let herself drop to the ground, into or close to the ongoing raid below. Here she once again elicits a recruitment response, and within minutes scores of army ants ascend into the rainforest understory in search of the promised spoils (Chadab and Rettenmeyer 1975).

Army ant recruitment is very fast and highly efficient compared to other ants (Chadab and Rettenmeyer 1975). While some ants draft via tandem running, in which the recruiter is being closely followed by one newcomer at a time, in army ants and some other ants the location of the prey is encoded in the recruitment trail instead. This means that many workers can be mobilized simultaneously. Furthermore, while the initial scout plays an important role in alerting other ants to the recruitment trail or the presence of prey in general, she is not required to lead the raiding party. Recruitment is therefore more continuous than in ponerine group raiders where the raiding party disbands if the scout is removed (see Chapter 2) (Chadab and Rettenmeyer 1975; Longhurst and Howse 1979; Mill 1984). Finally, because recruitment occurs locally at the raid front rather than the nest, the scout does not have to travel very far to solicit reinforcement (Chadab and Rettenmeyer 1975).

The *Eciton burchellii* swarm changes directions frequently in alternating flanking movements, and large swarms regularly split and reunite (Schneirla 1940; Franks 2001). Despite these dynamics, which mostly depend on the interactions between ants as well as local fluctuations in their densities, the swarm overall roughly maintains its compass bearing as long as the outpouring of recruits from the bivouac remains constant and strong (Schneirla 1934, 1940, 1971; Rettenmeyer 1963a; Rettenmeyer et al. 1983). It will only be later in the day, as the replenishment of workers weakens, that external factors such as prey availability and physical features of the local terrain gain a more dominant role in shaping the course of the raid (Schneirla 1940).

FIGURE 3.7 Army ants often form temporary booty caches along raiding columns. This picture shows a booty cache of *Eciton burchellii*, consisting mostly of looted ant brood.

In the wake of the swarm, the first clusters of ants are now covering arthropods that did not escape; as large items are chopped into pieces and the ants begin carrying their loot back to the nest, a fine network of fan columns emerges behind the swarm front. The ants temporarily deposit some of their booty in prey caches that often form at trail junctions (Figure 3.7). Ultimately, the fan columns coalesce into a single base column, three to twelve ants in width, which connects the swarm back to the bivouac and retraces the swarm's trajectory (Rettenmeyer 1963a; Schneirla 1971) (Figure 3.8).

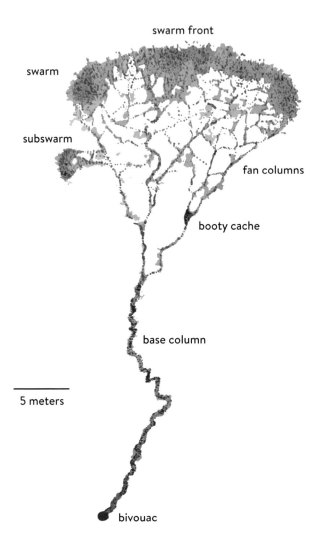

swarm front

swarm

subswarm

fan columns

booty cache

base column

5 meters

bivouac

FIGURE 3.8 Among army ants in the genus *Eciton*, *Eciton burchellii* has by far the largest colonies and conducts impressive swarm raids. The dimensions of such a carpet-like swarm can reach 20 meters across and 2 meters deep. In the wake of the swarm, ant traffic collapses into fan columns, which ultimately consolidate into a single base column that connects the swarm to the bivouac. The ants form booty caches where they intermittently store prey along the way. Reformatted from Rettenmeyer CW (1963a), "Behavioral Studies of Army Ants," *University of Kansas Science Bulletin* 44: 281–465, figure 6.

The more surface-adapted among the New World army ants tend to have expanded brain areas for visual processing compared with their more subterranean relatives, suggesting that simple optical cues like differences in light intensity play some role in these species (Werringloer 1932; Bulova et al. 2016). However, at least when it comes to seeing an actual image, all doryline army ants are thought to be essentially blind. Instead, they orient and communicate chiefly via tactile and chemical cues (e.g., Blum and Portocarrero 1964; Watkins 1964; Watkins et al. 1967a; Topoff et al. 1972a, 1972b, 1973; Topoff and Lawson 1979; Bagneres et al. 1991; Billen and Gobin 1996). The organization of the fan therefore principally arises from a combination of two things: increased congestion close to the swarm front that increases tactile interactions and diverts outbound ants, and the competition between fan columns for chemical persistence, which depends on differences in traffic flow and the associated reinforcement with trail pheromone (Schneirla 1940). As long as the swarm advances, the traffic on the base column flows in both directions, like cars on a highway, with booty-laden ants returning to the nest and outbound fighters on their way to the battlefield. While most army ant species are confined in their raiding activity to the soil or leaf litter, both the swarm raider *Eciton burchellii* and the column raider *Eciton hamatum* venture up into the vegetation, and especially *Eciton burchellii* raids can reach all the way into the canopy of giant rainforest trees (Teles da Silva 1982).

When traveling across steep slopes like tree trunks or over rugged terrain like the forest leaf litter, workers of both *Eciton burchellii* and *Eciton hamatum* assemble into living structures that facilitate traffic flow and help minimize energy expenditure at the colony level. Thomas Belt, who discovered the close mutualistic association between bullhorn acacias and needle ants discussed in Chapter 2, described some of the incredible self-assembling architecture of these army ants: "I once saw a wide column trying to pass along a crumbling, nearly perpendicular, slope. They would have got very slowly over it, and many of them would have fallen, but a number having secured their hold, and reaching to each other, remained stationary, and over them the main column passed. Another time they were crossing a watercourse along a small branch, not thicker than a goosequill. They widened this natural bridge to three times its width by a number of ants clinging to it and to each other on each side, over which the column passed three or four deep. Except for this expedient they would have had to pass over in single file, and treble the time would have been consumed. Can it not be contended that such insects are able to determine by reasoning powers which is the best way of doing a thing, and that their actions are guided by thought and reflection?" (Belt 1874, 27–28).

Some of the resulting infrastructure, such as the flanges described by Belt or the bridges that had been described earlier by Merian (see Chapter 1), seemed so wondrous that they were initially met with considerable skepticism (Figures 3.9 and 3.10; see also Figures 1.2 and 1.3). For example, the Reverend Lansdown Guilding, a natural historian studying in the Grenadines, who had never seen an *Eciton burchellii* colony, remarked in response to Merian's report that "one can scarcely credit their bridge of bodies, more singular than the chain-bridges of our rivers" (Guilding 1834, 363). However, although the descriptions by Merian and Belt were certainly accurate, we now understand that the sophisticated collective behaviors of army ants are not controlled by the "reasoning powers" of individuals but instead emerge from simple and local interaction rules.

The logic underlying army ant bridge formation in particular has received a considerable amount of recent attention (Powell and Franks 2007; Garnier et al. 2013; Reid et al. 2015; Graham et al. 2017). Ant bridges that span gaps in the leaf litter help create a paved roadway and shortcuts that increase traffic flow and therefore colony efficiency, both during raids and emigrations. Where the ants are traveling over relatively smooth terrain such as fallen tree trunks, no bridges at all might form over tens of meters of trail, while several bridges per meter can emerge wherever the ants are traversing uneven ground. Arguably the smallest "bridges" are pothole plugs, formed by single ants that stumble into a small gap and are then held in place by their sisters running overhead (Powell and Franks 2007) (Figure 3.11). True bridges can be composed of anywhere from a few workers that fill in a medium-sized crevice, to hundreds of workers forming a fully developed suspension bridge that measures many times the body length of an individual ant.

FIGURE 3.9 One of the most striking features of army ant biology is the construction of living bridges, as was already described in the first account of *Eciton burchellii* by Maria Sibylla Merian in 1705. The workers making up the bridge can remain in place for hours, as long as traffic continues. The upper panel shows a small group of *Eciton burchellii* workers collectively bridging a gap in the leaf litter. The lower panel shows other workers carrying prey across the same bridge.

FIGURE 3.10 Army ants are masters of living architecture. Here, *Eciton burchellii* workers are forming flanges at the intersection of two branches over which the raiding column is traveling. Flanges like these facilitate traffic flow by widening narrow surfaces and providing footholds on slippery and steep terrain. If you look carefully, you will spot a myrmecoid beetle, one of the army ant social parasites that will be discussed in Chapter 6, running on one of the flanges.

Large bridges originate in places where the ant trail strongly deviates from a straight path—for example, where the ants travel over two branches that cross at an angle (Reid et al. 2015). The bridge begins to form at the intersection, probably by a single ant serving as a flange. By a dynamic process of ants attaching to the far side and detaching from the near side, the bridge then extends out into the gap between the two branches to form an ever more efficient shortcut. The smaller the angle between the two branches and the heavier the traffic flow, the further the bridge moves as it lengthens and widens. As is the case for the living pothole plugs, the ants engaged in bridge formation are sensitive to the traffic flow overhead, and remain motionless as long as their backs are trod on. This simple interaction rule explains the movement of the bridge: because ants seeking the shortest path will tend to run across the far side, additional ants are likely to attach there and become immobile, while ants engaged in the near side of the bridge are more likely to leave the structure (Reid et al. 2015). And while the bridge stays in place even under small variations in traffic flow, it disassembles within seconds once traffic has stopped, based on the same local interaction rule (Garnier et al. 2013).

Another fascinating question is how traffic itself organizes on army ant trails. Here again, it turns out that simple local interaction rules are sufficient for traffic control. Both *Eciton burchellii* and the African driver ant *Dorylus molestus,* for example, employ trail pheromones produced in different types of specialized glands at the tip of the gaster (Billen 1992; Billen and Gobin 1996). When walking along trails during raids or emigrations, army ants have a tendency to move toward the highest concentration of trail pheromone in the center of the trail, resulting in compact and stable traffic columns. During emigrations, all ants are outbound from the old bivouac, while at the end of a raid all ants are returning either empty handed or laden with booty. However, during much of the day ants are returning to the nest with prey while others are simultaneously leaving the nest as reinforcement. Under bidirectional flow, ants should frequently face head-on collisions that force them to slow down their pace, congesting traffic and thus decreasing foraging efficiency. Yet army ants have evolved a simple trick to avoid severe traffic jams.

FIGURE 3.11 In its simplest form, an *Eciton burchellii* bridge consists of a single worker forming a pothole plug along the ants' path.

During collisions, outbound ants are more easily persuaded to change direction than inbound ants. This simple asymmetry during ant-ant interactions, in combination with the general attraction to the center of the trail, gives rise to a stable and efficient three-lane system in which returning ants occupy the central lane and outbound ants the outer two lanes (Couzin and Franks 2003). Lanes also emerge spontaneously in humans at high densities of pedestrians walking in opposite directions. Natural selection can act at the colony level to produce systematically asymmetric interactions in army ant collisions that ultimately translate into efficient lane formation, but humans are expected to behave selfishly in an effort to minimize their own travel time. Unlike in army ants, there is thus no reason to assume that the emergent patterns in humans would be adaptive at the group level (Couzin and Franks 2003).

Such collective behaviors, in which groups of individuals show emergent properties that are not apparent at the individual level, are common across biological systems, ranging from proteins and cells to bird flocks, fish shoals, and ant colonies. From the examples of raid and trail formation, to the living army ant bridges and flanges, the striking level of coordination is achieved via some simple local cue or set of cues that leads to large-scale correlations between individual behaviors. These mechanisms can give rise to group capabilities that far exceed the facilities of individuals, but the flipside is that the necessary underlying social conformism, when taken out of its natural context, can also produce detrimental behavioral inertia that the group is unable to escape.

The so-called army ant mills, or "death spirals," initially described by William Morton Wheeler, constitute a striking example of such inertia: "I have never seen a more astonishing exhibition of the limitations of instinct. For nearly two whole days these blind creatures, so dependent on the contact-odor sense of their antennae, kept palpating their uniformly smooth, odoriferous trail and the advancing bodies of the ants immediately preceding them, without perceiving that they were making no progress but only wasting their energies, till the spell was finally broken by some more venturesome members of the colony" (1910, 208). The phenomenon was later also observed by William Beebe (Beebe 1921) and studied in more detail by Theodore Schneirla (Schneirla 1944c) (Figure 3.12).

Death spirals typically emerge after a sizeable group of army ant workers has become cut off on featureless substrate, such as a paved road or sidewalk. This can happen when a raid or an emigration is interrupted by heavy rains. While the ants seek refuge under a large leaf, for example, the rain washes away the trail pheromone, and once the downpour has subsided and the surface has dried, the ants find themselves disconnected from the rest of their colony. As the ants begin to venture out from under their shelter, they lay trail pheromone on the pavement. However, the central cluster of ants provides a force of attraction that is

FIGURE 3.12 The collective behaviors of army ants can lead to nonadaptive phenomena, especially in artificial environments. In this image, *Eciton burchellii* workers walk in a circular mill or death spiral in a laboratory nest. These spirals can be quite persistent and sometimes only dissolve after many hours when the ants eventually start dying.

difficult to overcome, and the first timid scouts turn back after only a few centimeters. The result is a buildup of pheromone around the cluster, until a continuous circular trail has formed. More and more ants now begin to follow this trail until the entire cluster begins to spiral in the same direction. As the ants walk in a circle, they deposit more trail pheromone, reinforcing their march into oblivion often until they die of exhaustion.

Army ant death spirals are a somewhat artificial phenomenon that would soon be interrupted in the ants' natural habitat by features such as branches, roots, or simply the rugged contours of the leaf litter. So while ant mills constitute a misguided manifestation of the logic of army ant trail formation that emerges only in featureless man-made environments, they elegantly illustrate that the "intelligence" of an army ant colony is distributed across its members.

If *Eciton burchellii* is the archetypical swarm raider, *Eciton hamatum* is the best-studied column raider. In a column raid, the raid front is composed of small groups of ants that fan out at the tip of the column to cover an area of one or two square decimeters (Rettenmeyer 1963a; Schneirla 1971). There is a continuous turnover of ants at the raid front as some retreat while others push forward (see Chapter 2). During a fully developed raid, the column can be anywhere from one to six ants wide and over 300 meters in length (Rettenmeyer 1963a). In the Brazilian Amazon, the average raiding trail length of *Eciton hamatum* is 195 meters, ranging from 70 to 370 meters. In comparison, the average *Eciton burchellii* raiding trail length at the same site is only 75 meters, ranging from 20 to 280 meters (Teles da Silva 1982). And while an *Eciton burchellii* colony sends out a single swarm raid that remains connected to the bivouac via a base column, around three raiding columns can leave an *Eciton hamatum* bivouac at the same time in different directions (Rettenmeyer 1963a; Schneirla 1971) (Figure 3.13). Compared to an *Eciton burchellii* swarm raid, the daily prey intake of an *Eciton hamatum* column raid is of similar magnitude in terms of the number of pieces retrieved (Powell 2011).

The precise structure of an army ant raid, whether column or swarm raid, is once again thought to emerge from simple behavioral rules that

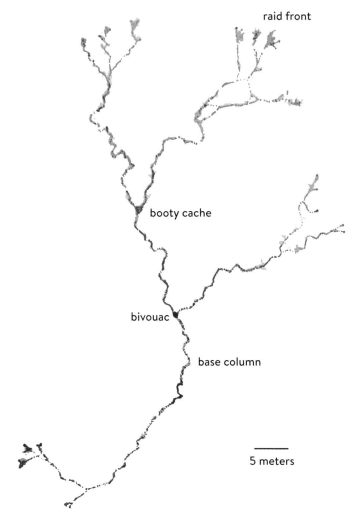

raid front

booty cache

bivouac

base column

5 meters

FIGURE 3.13 *Eciton hamatum* colonies are smaller than those of *Eciton burchellii*, their diet is limited to social insects, and they usually raid in columns. Several of these raiding columns can be active simultaneously. At the raid front, the ants spread out into miniature swarms, and prey is temporarily deposited in booty caches, often at intersections between main raiding columns. Reformatted from Rettenmeyer CW (1963a), "Behavioral Studies of Army Ants," *University of Kansas Science Bulletin* 44: 281–465, figure 5.

govern the interactions of the many foragers with each other and their environment (Deneubourg et al. 1989; Franks et al. 1991; Solé et al. 2000). One idea is that the spatial distribution of prey controls the structure of the emergent ant traffic and, therefore, the structure of the raid. Using computer modeling, researchers have compared the effects of two different types of prey distributions on the resulting raid structure. In the first, any spot in the virtual space was assigned a very low probability of containing prey. However, if prey was present, it was abundant, and there were many prey items available for harvest. This distribution can be thought of as approximating social insect populations, where resource-rich colonies are scattered sporadically across the landscape. In the second distribution, each spot had a high probability of containing food, but in each case only a single food item was present, a distribution that might be more representative of the prey of a generalist predator. Lo and behold, a column-like raid pattern emerged from the computer model based on the former prey distribution, while a swarm-like raid pattern emerged under the latter distribution.

In essence, spatially widely separated sites of prey capture translate into spatially separated ant trails and thus a raid that is split into discrete columns (Deneubourg et al. 1989). The researchers then followed up on their in silico study with a simple experiment in which they manipulated the prey distribution in front of an advancing *Eciton burchellii* swarm raid, either by creating bare patches on the forest floor by raking away the leaf litter, or by supplementing the natural resources by putting out a massive army ant care package in the form of 100 crickets inside a plastic mesh bag. As the swarm moved across the experimentally manipulated patch, it took on a more column raid-like pattern, just as the model had predicted (Franks et al. 1991). Additional theoretical work has since suggested that the specific local interactions underlying army ant raid organization might actually have been fine-tuned by natural selection and that the emergent properties of the raid thus not only constitute a response to the particular prey-type of any given species, but are also especially suited for the exploitation of that prey (Solé et al. 2000).

The spatial distribution of prey may have an effect on the emergent organization of the raid, but this is unlikely to be the whole story. After all, a typical *Eciton burchellii* raid sets out as a carpet swarm whereas an *Eciton hamatum* raid begins with small pushing parties that advance in columns (Schneirla 1933, 1934, 1940) (see Chapter 2). Similarly, at the beginning of a swarm raid, *Dorylus nigricans* driver ants form a dense sheet of excited workers at the nest exit. As the ants grow more and more belligerent, any external trigger—say, a few drops of water or a falling twig—is sufficient for this sheet to explode into a large swarm that begins to move away from the nest (Leroux 1977a, 1982). The early stages of *Labidus praedator* swarm raids also resemble those of *Eciton burchellii* (Schneirla 1933, 1934). *Aenictus* and *Eciton vagans* column raids, on the other hand, begin much more humbly, and very much like those of *Eciton hamatum* (Schneirla 1934; Schneirla and Reyes 1966). In other words, representative raids of swarm and column raiders differ in structure from the outset, before the ants have encountered any prey.

As we have discussed in Chapter 2, army ants probably underwent a gradual, emergent transition from scout-initiated group raiding to spontaneous mass raiding as colony size increased. A similar, gradual transition is also conceivable between column and swarm raiding. The first raids that qualified as spontaneous mass raids would have been weak column raids at intermediate colony sizes. As colony size increased further in some lineages, the flux of ants streaming to the raid front from the bivouac eventually became so large that, rather than forming a small, fan-shaped cluster spearheading a well-defined column, the ants began to spread into a dense, carpet-like swarm of ever increasing dimension. This transformation made it feasible to overwhelm large, mobile arthropods and therefore went hand in hand with an expansion in food spectrum. That the emergent architecture of the raid is indeed largely a matter of numbers becomes apparent in small raids of species that typically hunt in swarms or particularly large raids of species that typically hunt in columns. Theodore Schneirla (1933), for example, observed that "At times, however, when the *E. burchelli* raid does not begin as abruptly as usual, or when fewer ants participate, the advance itself is in the form of a smaller fan-shaped spreading, from which a narrow column extends to the bivouac. This is similar to the character-istic behavior of *E. hamatum*" (268). Conversely, during the peak of an

Eciton hamatum raid, the front can begin to resemble an *Eciton burchellii* swarm (Schneirla 1933).

Despite the fact that *Eciton burchellii* and *Eciton hamatum* workers differ in their behavior in ways that probably impact raid organization (Schneirla 1940), I propose that the local interactions in column and swarm raiders in fact need not be fundamentally different, but that the emergent properties of the raid are largely a matter of scale. This is of course speculative at this point, given that precisely how the number of participating ants affects raid organization has been explored neither empirically nor theoretically. Yet whether and how the properties of a collective system change qualitatively with the number of its constituents is an important problem of broad relevance. That the collective behavior of insect societies can undergo major transitions as a function of group size, even without major changes in the underlying behavioral rules, is thus a particularly intriguing idea (reviewed in Dornhaus et al. 2012). Raid organization in dorylines is certainly a fascinating natural phenomenon from which to delve into this question. Despite this great potential for identifying general rules underlying raid organization, we will see in the following sections that, on a species-by-species level, the army ants' superorganismal intelligence has been fine-tuned by natural selection to target rather specific prey types.

Social Insect Predators

Most army ants in the subfamily Dorylinae prey exclusively or almost exclusively on other ants (Rettenmeyer 1963a; Rettenmeyer et al. 1983; Gotwald 1995; Hoenle et al. 2019). This includes the vast majority of species in the genera *Aenictus* and *Neivamyrmex,* the two largest army ant genera in terms of species numbers (e.g., Gotwald 1976; Gotwald and Cunningham-van Someren 1976; Rettenmeyer et al. 1983; Rościszewski and Maschwitz 1994; Gotwald 1995; Hirosawa et al. 2000; LaPolla et al. 2002; Le Breton et al. 2007; Hashimoto and Yamane 2014; Hoenle et al. 2019). And even within the genus *Eciton,* most species are specialized ant predators. A systematic survey of prey collected from the six local

Eciton species and five other surface-raiding army ants at La Selva Biological Station, for example, found that 98 percent of the prey items were ants (Hoenle et al. 2019). In fact, 87 percent of prey items were ant brood of different developmental stages. Unfortunately, it is often impossible to identify ant brood to the species-level based on morphology, and it has thus been challenging to study which ant species exactly army ants hunt and to estimate how the different army ants in a given community divvy up the available prey.

A powerful complement to morphological identification has once again become available due to advances in molecular genetics: DNA barcoding. The idea is that a short fragment of DNA sequence, when compared with a well-annotated reference database, can reveal the species identity of its bearer. This not only allows researchers to match males and females of the same species, as alluded to in Chapter 1, but also assign species identities to different developmental stages. Using this approach, Philipp Hönle and colleagues (2019) found that the eleven studied army ant species had raided the nests of 129 prey ant species. In other words, even though their study was based on limited sample sizes and looked at only half of the army ant species present at La Selva, about 30 percent of the local ant fauna was represented among the collected prey. This suggests that, collectively, army ants raid the nests of the great majority of ant species in Neotropical rainforests. However, even though all army ants at La Selva take ant prey, each species is quite selective in their targets, and the overlap in diet between different army ant species is therefore limited. This high level of prey specialization is apparent at every taxonomic level, from the prey ant species, to the genus and subfamily (Powell and Franks 2006; Hoenle et al. 2019).

For example, both on Barro Colorado Island in Panama and at La Selva Biological Station in Costa Rica, *Eciton dulcium* attacks only colonies of ponerine ants, mostly in the genera *Odontomachus* and *Pachycondyla* (Powell and Franks 2006; Powell and Baker 2008; Hoenle et al. 2019). In *Eciton vagans,* on the other hand, prey ants also include for example myrmicines in the genera *Aphaenogaster* and *Pheidole* (Rettenmeyer et al. 1983; Powell and Franks 2006; Hoenle et al. 2019). Even the *Eciton* species with a broader palate still have clear prey preferences. On

FIGURE 3.14 Most doryline army ants, including those in the genus *Eciton*, prey predominantly or exclusively on other ants. Leaf-cutting ants of the genus *Acromyrmex*, for example, constitute a major dietary component of *Eciton hamatum*, both at La Selva Biological Station in Costa Rica (this picture) and on Barro Colorado Island in Panama. This picture was taken early into a colony emigration, when the ants are carrying both prey items from the day's raid (the two large *Acromyrmex* pupae) as well as their own larvae (in the foreground and background).

Barro Colorado Island and at La Selva Biological Station, *Acromyrmex* leaf-cutting ants constitute a large proportion of *Eciton hamatum* prey, while in Ecuador that species takes mostly formicine ants of the genera *Gigantiops* and *Camponotus* (Rettenmeyer et al. 1983; Powell and Franks 2006; Powell and Baker 2008; Powell 2011; Hoenle et al. 2019) (Figure 3.14). However, unlike the previously mentioned species, *Eciton hamatum*, at least in some populations, also regularly supplements its ant diet by plundering the nests of large social wasps, primarily in the subfamily Polistinae (Rettenmeyer 1963a; Rettenmeyer et al. 1983). The Amazonian species *Eciton rapax*, known mostly as an aficionado of *Camponotus*, *Odontomachus*, and *Pachycondyla* ants, is not averse to social wasps either (Rettenmeyer et al. 1983; Burton and Franks 1985). In South America, both *Eciton hamatum* and *Eciton rapax* have further been reported to attack nests of the Amazonian bumblebee *Bombus transversalis* (Ramírez and Cameron 2003). Finally, the most generalist species in the genus, the swarm raider *Eciton burchellii*, still derives approximately half of its diet from ants (Franks 1982a; Franks and Bossert 1983). On Barro Colorado Island, 96 percent of the retrieved prey ants belonged

FIGURE 3.15 Even though *Eciton burchellii* has the broadest prey spectrum among army ants in the genus *Eciton*, ants still constitute a large proportion of its diet. On Barro Colorado Island in Panama and at La Selva Biological Station in Costa Rica, carpenter ants in the genus *Camponotus* feature particularly prominently on the menu. This image shows an *Eciton burchellii* raid attacking the nest of an undescribed *Camponotus* species (*Camponotus* JTL005 on AntWeb.org) located inside the trunk of a large rainforest tree approximately 1 meter off the ground. The attack began at around 9:00 a.m., focusing on three large entrance holes to the nest that were blocked by the carpenter ants. It took the army ants over 1.5 hours until they gained access to the colony at one of the chosen points of attack. At around 10:45 a.m., the first carpenter ant workers were removed, and by 11:25 a.m., dozens of workers and alates had been dragged out of the various nest entrances (shown here). When I returned to the site at 4:00 p.m. that day, the army ants were still active on the tree, moving in and out of the carpenter ant nest, which at this point seemed to have been thoroughly plundered. At the base of the tree, a sizeable prey cache of *Camponotus* brood had formed, 10 centimeters from the main raiding trail leading back to the bivouac.

to the carpenter ant genus *Camponotus,* and *Camponotus* ants are also the preferred prey at La Selva (Powell and Franks 2006; Hoenle et al. 2019) (Figures 3.1 and 3.15). So in that sense even *Eciton burchellii* is quite specialized.

The dietary specialization of army ants might be mediated by a combination of factors. Experiments with *Eciton hamatum* showed that the ants are highly attracted to nest odors of prey ants but ignore nest material from other ants, suggesting that differences in chemosensory preferences could determine which ants a given army ant species recognizes as prey (Manubay and Powell 2020). Differences in general raiding behavior that determine which potential prey species a given army ant is most likely to encounter in the first place are probably also important. For example, while *Eciton dulcium, Eciton mexicanum,* and *Eciton vagans* mostly raid ants that nest in the soil and leaf litter, *Eciton burchellii* and *Eciton hamatum* have a much stronger tendency to attack arboreal ants, those that inhabit trees (Hoenle et al. 2019). Finally, overlap in raiding activity is further reduced by differences in activity patterns. *Eciton burchellii* and *Eciton hamatum* usually raid during the day, and *Eciton dulcium, Eciton mexicanum,* and *Eciton vagans* are mostly active at night (Hoenle et al. 2019). In summary, species in the genus *Eciton* alone display a broad range of diets and can be dietary specialists to vastly different degrees (Figures 3.16–3.18).

Just as different army ants specialize on attacking different prey ants, different prey species have evolved many different defensive responses.

FIGURE 3.16 Different species of army ants can differ widely in their prey spectra. Shown here is a sample of prey items collected within a few minutes from an *Eciton vagans* raid. The sample contains ant larvae, pupae, workers, males, and even a queen. *Eciton vagans* is a species with a rather narrow prey spectrum, and DNA barcoding and morphological identification showed that all items on this picture belong to a single species of trap-jaw ant, *Odontomachus panamensis.*

FIGURE 3.17 The species *Eciton hamatum* has a somewhat broader prey spectrum, including a larger range of ant species as well as social wasps. Here again, prey items were collected from a single raid within a few minutes to provide a snapshot of what the ants are bringing back to the bivouac. These include myrmicine ants of the genera *Acromyrmex* (most of the larger items, including all individuals in the top row and left two columns), *Pheidole*, and *Trachymyrmex* (the small pupae on the right), as well as formicine ants of the genus *Camponotus* (the small, more elongated larva in the bottom right corner) and ponerine ants of the genus *Neoponera* (the large, slender pupa in the center).

FIGURE 3.18 The picture changes dramatically when looking at prey items retrieved during an *Eciton burchellii* raid. Even though a large proportion of the captured booty are still ants, *Eciton burchellii* also takes a broad range of other arthropods. This small selection, collected within a few minutes from a single raiding column, includes, for example, the ants *Neoponera lineaticeps* (upper left), an undescribed *Camponotus* species (JTL21 on AntWeb.org; upper row, second from left), and *Pseudomyrmex* spec. (*gracilis* group; upper row, second from right). Among the non-ant prey are a centipede (family Scolopendridae), true bugs of the families Fulgoridae and Reduviidae, beetles of the families Carabidae and Scaraberiidae, as well as an earwig of the family Labiduridae.

In fact, it has been suggested that colony defense strategies and even worker caste morphology have in part evolved as a response to army ant predation pressure (e.g., LaMon and Topoff 1981). One behavior that can be readily observed in many ant species is nest evacuation, either immediately or after an initial period of active resistance, once the army ants gain access to the nest (McDonald and Topoff 1986). In the deserts of the southwestern United States, for example, entire colonies of the myrmicine *Novomessor cockerelli* and other ants vacate their nests in response to raids by the army ant *Neivamyrmex nigrescens* (LaMon and Topoff 1981; McDonald and Topoff 1986; Smith and Haight 2008). This response is so reliable that it provides a simple trick to collect entire mature colonies of *Novomessor cockerelli,* sparing researchers the sweaty task of excavating nests of several thousand ants from the hard desert soil. Within seconds of introducing a few hundred army ants into the entrance, a steady stream of workers will exit the nest; within about thirty seconds to a minute, the workers begin evacuating the brood. Soon the queen leaves as well, and the entire colony can be conveniently scooped up. It takes about eleven hours of hard work to collect a single *Novomessor cockerelli* colony via nest excavation, but the army ant trick does the job in just twenty minutes (Smith and Haight 2008).

Nest evacuations also frequently occur in response to *Eciton burchellii* and *Eciton hamatum* raids, with varying degrees of success. At La Selva Biological Station, workers of *Aphaenogaster araneoides* rush out of the nest carrying brood at the slightest disturbance, as if they were only waiting right at the nest entrance for army ants to attack (Figure 3.19). In several ant species, workers leave the nest as soon as the army ant approach is detected, carrying pupae and larvae in a partial evacuation of the colony. However, at least some of the brood is usually still pillaged (Dejean et al. 2013). In other species, the entire nest is evacuated, including the queen.

These evacuations can be triggered by alarm pheromones released by returning foragers that have encountered the army ants in a distance. The black crazy ant *Paratrechina longicornis,* for example, evacuates the nest long before the arrival of an *Eciton burchellii* swarm, forming a concentric cluster with queens in the middle, followed by a layer of

FIGURE 3.19 Different ant species have different defense tactics against army ants. A common approach is to evacuate the nest at the first signs of an attack. Here, a worker of *Aphaenogaster araneoides* (subfamily Myrmicinae) is holding onto a larva on the leaf litter while her colony is under siege.

workers carrying brood, and finally an outer layer of workers only. The outer workers frequently leave the cluster to zigzag around in the unnerving manner that has given the species the name "crazy ant." These workers arguably gather information about the current location of the attackers and then steer the cluster in the right direction away from the army ant swarm. This approach is often so successful that no sizeable amount of brood is lost (Dejean et al. 2013). Furthermore, when direct encounters are unavoidable, black crazy ants also fight back by spread-eagling *Eciton burchellii* workers and dousing them in formic acid (Figure 3.20). As noted previously, the behavior of army ants is largely guided by tactile and olfactory cues, and intimate physical contact with other ants that defend themselves can lead to the transfer of surface chemicals. This can cause confusion later on. After fights with *Pheidole megacephala,* for example, *Eciton burchellii* workers are sometimes killed by their nestmates (Dejean and Corbara 2014).

Social wasps also usually evacuate their nest when faced with army ants. When *Eciton burchellii* or *Eciton hamatum* raids approach a nest of the small wasp *Protopolybia exigua* attached to the underside of a leaf, the wasps can detect the ants from a considerable distance via visual and olfactory cues. They specifically respond to those two army ant species with synchronized fanning and buzzing. This alerts their nestmates of the imminent danger. All the adult wasps then exit the nest via the single small entrance in single file and spread out on the nest surface. As soon as the first army ants actually run onto the nest, the adult wasps fly off in unison. While the brood and often a few adult wasps that have remained inside the nest are lost to the ants, the casualties would arguably be much higher without the advance warning that allows the wasps to evacuate the nest in an orderly manner and prepare for a timely departure. Once the army ants have moved on, the wasps can return home or resettle elsewhere (Chadab 1979). Similar alarm and evacuation responses to army ants are known from several other wasps (Jeanne 1970; Young 1979; Chadab-Crepet and Rettenmeyer 1982; O'Donnell and Jeanne 1990). Some larger wasp species also grab the first army ants that attempt to enter the nest with their mandibles, fly away some distance from the nest, and drop the ants. This at least delays further ant recruitment to

FIGURE 3.20 Some ants fight back when army ants approach. This picture shows black crazy ants (*Paratrechina longicornis*; subfamily Formicinae; also called the longhorn crazy ant) spread-eagling a much larger *Eciton burchellii* worker. The liquid drop on the army ant's head is probably formic acid, the black crazy ants' weapon of choice. The black crazy ant most likely hails from Africa and has become one of the world's most successful invasive species, now being considered a worldwide pest. In the tropics and subtropics, it is frequently found in gardens, parks, and other disturbed habitat, and it also lives inside buildings at higher latitudes. This scene at La Selva Biological Station was photographed right behind one of the research buildings.

the wasp nest and, if the raiding trail is sufficiently weak, might prevent a full-blown attack altogether (Chadab-Crepet and Rettenmeyer 1982). Other wasps are able to detect and remove ant trail pheromone, thereby diverting approaching ants away from their nest (West-Eberhard 1989).

A peculiar yet efficient response has been reported for the ant *Azteca andreae,* one of the species associated with *Cecropia* trees (see Figures 2.9 and 2.10). As an *Eciton burchellii* swarm approaches, the *Azteca* workers drop a few brood items from their nest entrance onto the ground at the base of their tree. And as the army ants gather the offerings, the actual swarm moves on instead of ascending the stem. The *Azteca* literally sacrifice some of their babies to avert greater harm to their society (Dejean et al. 2013). Other plant ants of the genera *Azteca* and *Pseudomyrmex* seem to have a repulsive effect on *Eciton burchellii* and *Eciton hamatum* that is possibly mediated via chemicals the ants deposit on their host tree (Dejean et al. 2013). Notably, several social wasps that are themselves potential victims of army ants frequently build their nests close to the colonies of aggressive arboreal ants, particularly on ant plants. By doing so, they seem to take advantage of the repellent effect of the resident ants on army ants (Chadab-Crepet and Rettenmeyer 1982; Dejean et al. 2001; Corbara et al. 2018).

Some ant species, like *Pachycondyla villosa* or *Ectatomma ruidum,* have small nest entrances that can be successfully plugged by one or a few workers to prevent army ants from entering (Dejean et al. 2013). The same behavior is displayed by *Pheidole obtusospinosa* in Arizona. This species possesses particularly large "supersoldiers," which alternate between blocking the entrance with their enormous heads and engaging in direct combat with attacking *Neivamyrmex* army ants outside the nest (Huang 2010). In other species, the nest architecture shows sophisticated modifications that have been interpreted as specific adaptations to dealing with army ants. At La Selva Biological Station, for example, two species of the genus *Stenamma* build intricate nests on nearly vertical clay banks inside the rainforest, with a pedestal-like structure serving as an elevated nest entrance (Longino 2005). Jack Longino, who has spearheaded an incredibly detailed survey of the ants of La Selva Biological Station and discovered the peculiar *Stenamma* nests, has suggested the plausible possibility that the elevated nest entrance serves to decrease the probability that it will be encountered by an army ant swarm or raiding column in the first place, and that it could also serve to dissipate the colony's odor, further impeding detection (Longino 2005). Interestingly, mature colonies actively maintain two or even three functional nests in close proximity, while actually occupying only one of those nests at any given time. Frequent colony relocations between nearby nests could be yet another mechanism to minimize the accumulation of colony odors that would otherwise attract army ants, and the same has been suggested for *Aphaenogaster araneoides* (McGlynn et al. 2004; Longino 2005). Efficient evacuations to alternative nests could also be specifically triggered by army ant attack (Droual 1984). Finally, the ants keep a spherical pebble right next to the entrance to each nest and replace this pebble within minutes if it is removed. If an army ant presents itself near the nest, a *Stenamma* worker will immediately grab the pebble and pull it inward like a stopper, thereby blocking the nest entrance (Longino 2005). It seems as though *Stenamma* has come up with a whole suite of complementary approaches to keep army ants at bay. Similarly, different aspects of wasp nest architecture, such as envelopes around the combs or a thin stem, or petiole, from which the comb is suspended, have been interpreted as adaptations to ant attack (Jeanne 1975).

Colonies of other species, such as those of the leaf-cutting ant *Atta cephalotes,* are simply avoided by *Eciton burchellii* (Dejean et al. 2013). And even where trails of the two collide, aggression does not escalate beyond a few sparring matches, possibly because the losses on both sides would be too severe (Figure 3.21). However, the story is an entirely different one when it comes to the epic battles between the mostly subterranean yet huge colonies of *Nomamyrmex esenbeckii* army ants and those of *Atta* leaf-cutting ants (Swartz 1998; Sáchez-Peña and Mueller 2002; Powell and Clark 2004). In fact, *Nomamyrmex esenbeckii* is the only known predator to successfully attack and even kill mature *Atta* colonies, which can be many million workers strong (Powell and Clark 2004). Upon attack, the leaf-cutters rapidly recruit their large soldiers to the frontline of battle in an effort to halt the onslaught and prevent the army ants from entering the nest. These fights can extend over an area of

FIGURE 3.21 The army ant *Eciton burchellii* and the leaf-cutting ant *Atta cephalotes* are arguably the two superstars of Neotropical myrmecology. Where the army ants' raiding or emigration columns collide with the foraging activity of the leaf-cutting ants, small scuffles can arise. This image shows two *Eciton burchellii* workers taking on a huge *Atta cephalotes* soldier, grabbing her by the antennae.

30 meters and last for a couple of hours. The leaf-cutters also respond by barricading their nest entrances with soil. Frequently, this tactic is successful, and the attackers eventually retreat. But if the army ants gain entry and conquer the nest, the consequences for the leaf-cutters are catastrophic. In one such case, *Nomamyrmex esenbeckii* workers removed approximately 60,000 brood items from an *Atta cephalotes* colony over the course of thirty-six hours, totaling over 1 kilogram of loot (Powell and Clark 2004). Once the army ants had moved on, the main battlefield on the nest surface, about 3 meters long and 0.5 meters wide, was littered with hundreds of dead *Atta* soldiers. A few days later, the colony was found dead (Powell and Clark 2004).

Next to ants, termites are the most prominent social insects in the tropics. Maybe unsurprisingly then, some army ants have homed in on termites. Army ant species of the *Dorylus* subgenus *Typhlopone,* for example, feed predominantly on termites, sometimes killing off entire mounds (Bodot 1961; Darlington 1985). However, these battles are difficult to observe because the ants enter the termite fortresses from below ground (Schöning and Moffett 2007). Several other army ants attack termite colonies at least occasionally, and some might even be quite specialized in that respect (e.g., Gotwald 1995; Schöning and Moffett 2007; Powell and Baker 2008; Souza and Moura 2008).

As we have seen in this section, most army ants prey almost exclusively on other social insects. However, the few swarm-raiding species with massive colonies and broad diets are those that stimulate human imagination the most and regularly make it into nature documentaries and popular culture.

Generalist Predators

Among the army ants with the most generalist diet are the swarm raiders *Eciton burchellii* and *Labidus praedator* in the Neotropics, as well as the African driver ants of the genus *Dorylus*. The mandibles of African driver ants have sharp cutting edges and overlap when fully closed, but the mandibles of *Eciton burchellii* and *Labidus praedator,* with the exception

FIGURE 3.22 While the mandibles of African *Dorylus* driver ants function like scissors with sharp blades that can cut through flesh, *Eciton* army ants employ their mandibles more like pliers when dismembering prey. *Eciton burchellii* at Alta Floresta, Mato Grosso, Brazil.

of the hook-shaped mandibles of the *Eciton* soldiers, of course, have rather blunt and poorly toothed edges that abut (Rettenmeyer 1963a). As a result, the mandibles of African driver ants function like scissors, allowing the ants to strip flesh even from larger vertebrates, while the mandibles of *Eciton burchellii* and *Labidus praedator* work more like pliers. Unlike *Dorylus* army ants, however, New World army ants sting their prey, injecting potent venoms that, in *Eciton burchellii* at least, have high activity of proteases, enzymes that break down proteins (Schmidt et al. 1986). Once the prey has been dissolved from the inside, *Eciton burchellii* and *Labidus praedator* butcher large arthropods by simply pulling and clipping them into pieces (Powell and Franks 2005) (Figure 3.22). The mandibles of *Eciton* and *Dorylus* soldiers are in fact so strong that indigenous peoples have traditionally used them to stitch wounds (Gudger 1925; Ramos-Elorduy de Concini and Pino Moreno 1988; Seignobos et al. 1996) (Figures 3.23 and 3.24). A series of biting

FIGURE 3.23 Being bitten and stung by an *Eciton burchellii* army ant is not a particularly pleasant experience. Here, a soldier is piercing the skin of the photographer's finger with her hook-shaped mandibles, while simultaneously using her mandibular foothold as leverage to drive in the stinger on her rear end to inject venom.

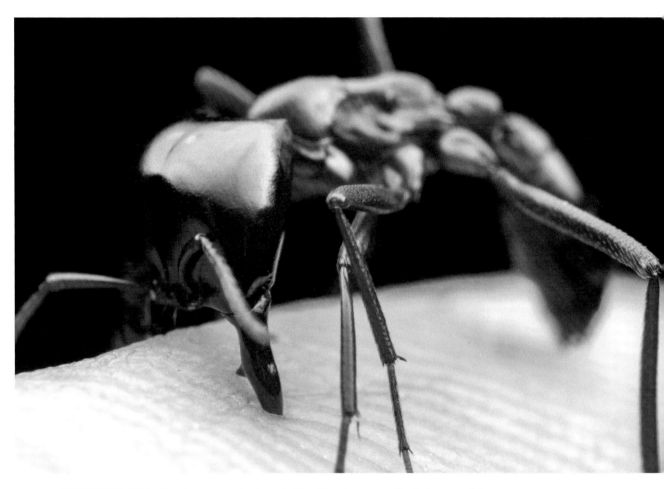

FIGURE 3.24 While *Dorylus* army ants do not sting, they have a particularly powerful bite. Here, a large *Dorylus wilverthi* worker is taking a stab at the photographer's finger. Kakamega Forest, Kenya.

ants are lined up in a row over the wound, with the tension of the mandibles keeping the wound closed. The bodies of the ants are then snapped off, with only the heads and mandibles remaining attached, forming a reliable suture. The *Eciton* soldiers with their peculiar mandibular morphology, however, are entirely specialized on colony defense, and do not partake in prey processing.

Although *Eciton burchellii* and *Labidus praedator* sometimes will kill a small vertebrate such as a lizard, frog, or nestling bird, they are generally believed to be unable to efficiently dismember and consume it (Schneirla 1934, 1971; Weber 1941; Rettenmeyer 1963a; Powell and Baker 2008). Nevertheless, there are occasional reports of *Eciton burchellii* indeed killing and stripping small vertebrates such as *Anolis* lizards or possibly even snakes to the bones (Rettenmeyer 1963a; Teles da Silva 1982; Sazima 2015, 2017). Carl Rettenmeyer also reported that he repeatedly saw lizard tails being carried in the raid columns of *Eciton burchellii*. The tails had probably been shed by the lizard in an effort to escape, a common defense tactic among lizards known as autotomy. Either way, these events are clearly rare exceptions rather than the rule, but possibly constitute the source of some of the exaggerated stories discussed in the Prologue.

The only regular vertebrate predator among the New World army ants might be the poorly studied genus *Cheliomyrmex*. *Cheliomyrmex* raids are usually entirely subterranean and have therefore been rarely encountered. However, the mandible morphology of *Cheliomyrmex* shows functional resemblance to that of African driver ants. On one occasion, researchers were lucky enough to observe workers from a colony in Ecuador stripping meat off a dead snake that the ants had possibly killed (O'Donnell et al. 2005).

Eciton burchellii swarm raids frequently go after wolf spiders and even the largest tarantulas, however. Amblypygids and scorpions, large and small, are also a staple on the menu (Figure 3.25). Among the additional items found most frequently carried in raid columns are pieces of crickets, katydids, cockroaches, and centipedes (Rettenmeyer 1963a; Schneirla 1971; Teles da Silva 1982; Vieira and Höfer 1994; Gotwald 1995). Army ants are extremely responsive to tactile cues from movements and vibrations, and a struggling large prey animal gets covered by

FIGURE 3.25 *Eciton burchellii* has by far the broadest diet among species in the genus *Eciton*, frequently capturing large and bulky prey. Here, a submajor and a regular worker have teamed up to carry a small scorpion back to the nest.

anywhere from fifty to over 200 *Eciton burchellii* workers in a matter of seconds (Teles da Silva 1982) (see Figure 3.6). At that point, of course, there is no escaping.

But this deadly attraction to action can also be exploited. Some beetles, for example, simply retract their legs and remain motionless when faced with an army ant swarm, thereby avoiding the fate that a more jumpy insect would face (Rettenmeyer 1963a). Some walking sticks also remain unfazed and emerge from the swarm unharmed (Otis et al. 1986). Thomas Belt (1874) made a similar observation with respect to "a green, leaf-like locust": "This insect stood immovably amongst a host of ants, many of which ran over its legs, without ever discovering there was food within their reach. So fixed was its instinctive knowledge that its safety depended on its immovability, that it allowed me to pick it up and replace it amongst the ants without making a single effort to escape" (19).

Other arthropods manage to get out of harm's way by employing fairly simple tricks. Some spiders and caterpillars, for example, suspend themselves from the plant they were just resting on with a silken thread as soon as the ants approach (Belt 1874; Rettenmeyer 1963a; Gotwald 1972). Harvestmen have yet another skillful maneuver up their sleeve, as once again described by Thomas Belt (1874): "I once saw one of the false spiders, or harvest-men (Phalangidae), standing in the midst of an army of ants, and with the greatest circumspection and coolness lifting, one after the other, its long legs, which supported its body above their reach. Sometimes as many as five out of its eight legs would be lifted at once, and whenever an ant approached one of those on which it stood, there was always a clear space within reach to put down another, so as to be able to hold up the threatened one out of danger" (19). Some insects, such as certain Hemiptera, or true bugs, seem to produce defensive chemicals that deter the ants (Rettenmeyer 1963a). Finally, in dung beetles (Scarabaeidae), which simply keep going about their putrid business in the midst of army ant swarms, it is not clear why the ants do not attack (Emlen 1996; Krell 1999).

Arguably the most generalist diets among New World army ants can be found in the genus *Labidus.* The two best-studied species are *Labidus praedator* and *Labidus coecus.* Both are largely subterranean in their nesting, emigration, and raiding activity. However, *Labidus praedator* sometimes forms carpet swarms on the forest floor that can rival those of *Eciton burchellii* in size although these are usually rather short-lived (Rettenmeyer 1963a; Powell and Baker 2008). Where the emigration or foraging columns of the two species are forced to surface, they are typically covered by tunnels that the ants construct out of soil particles. Similar to *Eciton burchellii,* both species prey on ants and other social insects as well as a wide range of solitary arthropods. However, unlike *Eciton burchellii, Labidus praedator* and *Labidus coecus* also take plant matter including seeds, nuts, and fruits, and *Labidus coecus* is easily attracted to oil baits (Borgmeier 1955; Rettenmeyer 1963a; Schneirla 1971; Fowler 1977; Perfecto 1992; Vieira and Höfer 1994; Gotwald 1995; Monteiro et al. 2008; Powell and Baker 2008).

In Africa, driver ants of the genus *Dorylus* form massive swarm raids composed of up to several million workers, surpassing even the raids of

Eciton burchellii and *Labidus praedator* in the New World in terms of size. Just like *Labidus,* and unlike *Eciton, Dorylus* driver ants nest in subterranean cavities, and where their columns surface the ants usually construct earthen tunnels or half-pipes (Figure 3.26). In a detailed study of the prey spectrum of the driver ant *Dorylus molestus* at Mount Kenya, Caspar Schöning and colleagues (2008c) found that the ants' diet largely consisted of earthworms and diverse insects, but also included arachnids, collembolans, isopods, land crabs, large slugs, centipedes, millipedes, large animal carcasses, and even plant seeds (Figures 3.27–3.29). Social insects, on the other hand, were sparse among the prey. Other ants constituted a very small proportion of the diet, and social wasps and bees were missing entirely. The only notable social insect prey items were the seasonally available dispersing young queens and kings of the termite *Odontotermes montanus* (Schöning et al. 2008c). Even though the exact diets of driver ants can vary substantially between different species, and even seasonally and between populations of the same species, this kind of very broad diet is rather typical (e.g., Gotwald 1974, 1995; Schöning et al. 2008c).

The fictional descriptions from the Prologue appear at least somewhat less exaggerated when applied to African driver ants. For example, the British herpetologist Arthur Loveridge (1922) reported that in the course of his battle with a colony that besieged his house in Tanzania for five entire days, several reptiles, including a chameleon and a caged crocodile from his collection, were stripped to the bones by the ants. Likewise, Bill Gotwald, an expert on African army ants, has repeatedly observed driver ants attacking, killing, and eating caged snakes (Gotwald 1984–1985). Recapitulating the horrors of García Márquez's *One Hundred Years of Solitude* (1970), Gotwald was told by a Ghanaian village chief that a baby had been killed by a swarm of driver ants when it was left unattended beneath a tree while the mother was tilling the garden (Gotwald 1984–1985, 1995). The wrath of the *siafu* also struck Jonathan, the one-year-old son of famed paleoanthropologists Louis and Mary Leakey, who were researching human origins in Kenya in 1942: "One evening, we heard a most awful wail from Jonathan, and we both rushed into his bedroom and switched on the light. He was sleeping in a cot

FIGURE 3.26 Army ants with more subterranean lifestyles move a lot of dirt, both during nest and trail construction. This must have important yet poorly understood effects on soil aeration and chemistry. Here, a colony of the African driver ant *Dorylus molestus* has constructed an earthen half-pipe around its trail as it crosses the forest floor at Mount Kenya, Kenya.

FIGURE 3.27 Unlike most other army ants, species in the Old World genus *Dorylus* are not specialized ant predators. In the African driver ant *Dorylus wilverthi*, for example, earthworms constitute a significant proportion of the diet. Kakamega Forest, Kenya.

FIGURE 3.28 The raids of some army ants, including *Eciton burchellii*, *Eciton hamatum*, and African *Dorylus* driver ants, ascend high up into the vegetation. Here, *Dorylus molestus* is attacking a large slug (possibly *Trichotoxon heynemanni*) in the rainforest of Mount Kenya, Kenya. The ants encountered the slug at around 11:30 a.m. during a large swarm raid. Initially, the slug, which was resting on a twig at a height of about 1.70 meters, hunched into a compact shape and secreted a lot of mucus, forming an envelope in which many of the ants got stuck. The slug persevered in this manner for about 1.5 hours, while a few hundred ants continued to slowly cut through the layer of slime. At about 1:05 p.m., being no longer able to withstand the ants, the slug emerged from its capsule and attempted to escape, with several ants still attached (shown here). After advancing for a few centimeters and leaving behind a thick trail of mucus, the slug detached from the twig and dropped to the forest floor, where it was immediately engulfed by driver ants (see Figure 3.29).

FIGURE 3.29 *Dorylus molestus* driver ants are butchering a slug (possibly *Trichotoxon heynemanni*) on the rainforest floor at Mount Kenya, Kenya.

under a mosquito net, and we had thought him safe from any kind of attack by insects. Unfortunately, we had overlooked the fact that siafu will climb up the legs of furniture and find a means of penetrating almost anything! We found, to our horror, that Jonathan was completely covered with hundreds of these biting ants. They were already all over his cot, and the floor was a seething mass of insects" (Leakey 1974, 157). But not only the Leakeys' son had to suffer from the siafu: "These ants are a real scourge to domestic animals and livestock in Africa. They attack puppies and kittens—who are relatively helpless when they are born—sitting hens, ducks, and turkeys, and even lambs and calves. The suddenness of the attack by thousands of ants all biting simultaneously makes it practically impossible for the mother to remove her young in time to save their lives" (Leakey 1974, 158). So while at least African driver ants can attack, kill, and devour larger vertebrates, this usually happens when the animals are caged or otherwise unable to escape.

With the possible exception of the termite predators of the *Dorylus* subgenus *Typhlopone* discussed earlier, army ants in the genus *Dorylus* in general have broad diets. For example, similar to driver ants, the subterranean Asian species *Dorylus laevigatus* feeds on earthworms, termites, some ants, isopods, beetles, grasshoppers, caterpillars, and even plant oils, which, as for *Labidus coecus,* can be used as efficient baits to attract that species (Weissflog et al. 2000; Berghoff et al. 2002a, 2002b). Arguably the army ant with the most highly developed vegetarian inclinations, however, is *Dorylus orientalis,* which is regularly caught feeding on a wide variety of cash crops and garden vegetables and has thus acquired a reputation as an agricultural pest (Gotwald 1995; Niu et al. 2010). Finally, swarm raiding in combination with generalist diets can also be found in ponerine army ants of the genus *Leptogenys* as well as in the marauder ants of the myrmicine genus *Carebara* (Moffett 1984, 1988a, 1988b; Maschwitz et al. 1989).

Even though army ants that prey on large arthropods dissect their prey before transport, the resulting pieces, such as the long legs of katydids or the stingers of large scorpions, are often still too bulky to be carried by a single ant. Remarkably, these species therefore form teams to achieve the task at hand. In *Eciton burchellii,* simple prey retrieval teams are usually composed of a submajor as the front runner piloting the effort, along with a regular worker lifting up the back end of the prey item to prevent it from dragging on the ground (Franks 1986; Franks et al. 1999; Franks et al. 2001) (see Figures 3.9 and 3.25). Persistent friction between prey and substrate, and the resulting drop in speed, can drive additional dynamic changes in team composition, and larger teams including several submajors can form when prey items are especially inconveniently shaped (Powell and Franks 2005) (Figure 3.30). Army

FIGURE 3.30 Army ants often transport large prey items collectively, and *Eciton burchellii* even has a specialized porter caste for this task, the submajor. Here, a group of submajors are schlepping a giant hind femur of a katydid (subfamily Pseudophyllinae) back to their bivouac.

ant teams are superefficient in that they can carry prey that is heavier than the sum of the maximum weights that individual team members could carry. This appears to be true for both *Eciton burchellii* and African *Dorylus* driver ants (Franks 1986; Franks et al. 1999; Franks et al. 2001).

Notably, *Eciton* submajors constitute a morphologically highly specialized porter caste. Their disproportionally long hind legs allow them to push themselves sufficiently off the ground to carry even large items in the army ant–typical manner: slung underneath the body (Topoff 1971; Powell and Franks 2005). This form of transport is energetically efficient, given that loads carried by individual army ants are typically lightweight and compact, compared, for example, to the large pieces of plant material that leaf-cutting ants carry above the head (Bartholomew et al. 1988; Feener et al. 1988). Accordingly, the submajor caste of *Eciton burchellii,* the only *Eciton* species that regularly takes large solitary arthropods that have to be butchered into bulky pieces for transport, is particularly well developed (Powell and Franks 2006). Evidently, army ants have evolved many behavioral and morphological adaptations to their predatory lifestyle. However, their purely carnivorous diet might pose an additional challenge: being left with a heavy stomach.

Due to advances in DNA sequencing technology, it has become clear over the past fifteen years or so that many animals, including humans, rely on microbial communities inhabiting the intestines for healthy digestion, nutrient uptake, and immunity. Given their specialized diet, this might also apply to army ants. And indeed, in the New World genera *Eciton, Labidus,* and *Nomamyrmex,* two types of seemingly specialized microbes, the Firmicutes and the Entomoplasmatales, dominate the gut microbiome. These microbes also occur in other army ant genera at lower frequencies, including Old World army ants, suggesting that this relationship, like the predatory habits of the dorylines, has persisted for many millions of years (Funaro et al. 2011; Łukasik et al. 2017). What role exactly these microbes play in the biology of army ants is currently unknown, but given the specific and ancient nature of the association, it is likely an important one, possibly related to processing the ants' meat-heavy diet.

Swarm Followers

The swarm raid of an *Eciton burchellii* colony is arguably among the most impressive spectacles the natural world has to offer, both optically and acoustically. As the swarm approaches, the crackling of a million tiny feet closing in on fleeing insects that scurry through the leaf litter sounds like a constant drizzle of rain. And while the rainforest usually is a hushed place during the day, the front of an army ant swarm raid is alive with vocalizing birds hopping up and down tree trunks and fluttering back and forth. Likewise, parasitoid flies buzz around the area ahead of the swarm, some of them intermittently perched on leaves and branches, others hovering over the ants like miniature "elfin helicopters" (Gotwald 1995). But neither the birds nor the flies are here for the ants themselves—they are after the escaping roaches, grasshoppers, and spiders. As Thomas Belt put it, they "accompany the ants, as vultures follow the armies of the East" (1874, 21). As we will see in this section, for the fleeing insects this means out of the frying pan and into the fire.

In the Neotropics, about fifty species of birds regularly attend swarm raids of *Eciton burchellii* and *Labidus praedator,* and many additional birds visit ant swarms occasionally (e.g., Gochfeld and Tudor 1978; Willis and Oniki 1978; Wiley 1980; Coates-Estrada and Estrada 1989; Dobbs and Martin 1998; Cody 2000; Vallely 2001; Rettenmeyer et al. 2011; O'Donnell 2017). Other New World army ants do not form swarm raids above ground, and they therefore usually do not attract birds. Observations of small antbird assemblages at *Eciton hamatum* raids, for example, seem to be an exception (Tórrez et al. 2009). However, the phenomenon also occurs at the raids of African driver ants (Peters et al. 2008; Peters and Okalo 2009; Schöning et al. 2010). In the Neotropics, large army ant swarms can attract well over twenty individual birds from up to twenty different species at a time (Willis and Oniki 1978). Dominance hierarchies exist between different bird species, with some occupying the preferred spots right above the swarm front while keeping other species confined to peripheral, less yielding zones (Willis and Oniki 1978).

The area around the swarm is further divvied up by perch type. Wood-creepers, for example, climb directly on large tree trunks in an upright, vertical posture, while most antbirds cling to slender vertical stems or horizontal branches (Willis and Oniki 1978). The ant-following birds do not purposefully feed on the army ants directly, but they nevertheless have a negative impact on army ant prey intake by snatching some of the arthropods the ants have either already pinned down or would otherwise have captured a few seconds later (Willis and Oniki 1978; Chesser 1995; Wrege et al. 2005).

At La Selva Biological Station, twelve species of birds attend more than 10 percent of swarm raids, including several species of woodcreepers, motmots, antbirds, and tinamous (Chaves-Campos 2003). One species, the ocellated antbird *Phaenostictus mcleannani,* has a particularly close association with the ants and can be seen at nearly all swarm raids at La Selva (Chaves-Campos 2003) (Figure 3.31). Like several other species, it is an obligate ant follower; in other words, its survival and reproduction depend on the presence of the ants. Obligate ant followers track the position of several of the frequently moving army ant colonies simultaneously, and inspect the bivouacs in the morning before following the ant trail to the swarm front (Willis and Oniki 1978; Swartz 2001; Chaves-Campos 2003). But even among the obligate ant followers the ocellated antbird stands out.

The ocellated antbird checks bivouacs more frequently than any other species and is represented in the highest numbers at ant swarms where it essentially obtains all its food (Chaves-Campos 2003). It is also dominant over other smaller species of antbirds and occupies the best perches right above the center of the swarm (Willis 1973). It is usually the first bird species to arrive at swarms, and its vocalizations then guide additional species of ant-following birds to the feast (Chaves-Campos 2003; Pollock et al. 2017). Like other obligate ant-following birds, ocellated antbirds keep track of several *Eciton burchellii* colonies within their feeding range. They sometimes vocalize right before leaving a swarm raid and flying on to the next ant colony. Other birds of the same species then follow along, forming a cohesive group that travels through the forest together. In other words, the birds seem to share a collective knowledge of where in the forest the different army ant colonies are currently hunting (Chaves-Campos 2011). The birds themselves are

FIGURE 3.31 The swarm raids of *Eciton burchellii* are accompanied by specialized birds that feed on the fleeing insects flushed by the ants. Here, an ocellated antbird (*Phaenostictus mcleannani*) is perched above the swarm front on the lookout for an easy meal. The ants are rushing up the light-colored buttress root of the large tree in the center of the image.

FIGURE 3.32 Several species of butterflies regularly attend *Eciton burchellii* swarm raids, where they feed on the fresh droppings of ant-following birds. This picture shows *Melinaea lilis imitata*, one of three Ithomiini butterflies that are common swarm followers at La Selva Biological Station (Ray and Andrews 1980).

certainly the most intensively studied swarm attendants (reviewed in Gotwald 1995; for additional, more recent studies, see Roberts et al. 2000a; Willson 2004; Brumfield et al. 2007; Chaves-Campos and DeWoody 2008; O'Donnell et al. 2010, 2012, 2014; Logan et al. 2011; Driver et al. 2018; Martínez et al. 2018). However, their presence in turn attracts additional visitors.

Certain species of butterflies, for example, seek out army ant swarms to feed on the nutrient-rich fresh bird droppings, an otherwise rare and distributed commodity in tropical rainforests (Ray and Andrews 1980; Rettenmeyer et al. 2011). In particular, female butterflies require nitrogenous compounds to produce eggs, and these compounds are scarce in flower nectar, but abundant in bird poop. It is thus mostly the female butterflies that are attracted to ant swarms (Ray and Andrews 1980). At La Selva, mostly three species of butterflies in the tribe Ithomiini accompany the assemblages of ant-following birds (Ray and Andrews 1980) (Figure 3.32).

Interestingly, it has been noted that, in African rainforests, ant-following birds are particularly likely to carry blood-born parasites such as *Trypanosoma* and *Plasmodium,* which are related to the parasites causing sleeping sickness and malaria in humans, respectively (Peters 2010). These parasites are usually transmitted via different types of flies, such as tsetse flies and mosquitoes in the case of our own parasites. Although this is pure speculation, the possibility that some flies might seek out army ant swarms to feast on ant-following birds is intriguing.

However, most of the flies you will encounter at an army ant swarm are not after the birds but the arthropods. In fact, the second large group of regular swarm attendees consists of these types of parasitic flies. Even though they have received considerably less attention than the ant-following birds, they are even more numerous—possibly numbering in the thousands at any given *Eciton burchellii* raid (Rettenmeyer et al. 2011). These flies belong to a number of different families, most notably thick-headed flies (Conopidae), tachinid flies (Tachinidae), scuttle flies (Phoridae), and flesh flies (Sarcophagidae). Both male and female thick-headed flies of the genus *Stylogaster* frequently hover over the swarm front, zipping to and fro, or mating at the periphery of the swarm

FIGURE 3.33 Scores of parasitoid flies accompany the swarm raids of *Eciton burchellii* and *Labidus praedator*. These flies deposit their eggs or larvae into the arthropods that the ants flush from the leaf litter. Here, a conopid fly of the genus *Stylogaster* is hovering above a *Labidus praedator* swarm raid in Henri Pittier National Park, Venezuela. *Stylogaster* females dive down to jab their harpoon-shaped eggs into the host. The *Stylogaster* larvae then develop as endoparasitoids inside the host insect.

FIGURE 3.34 Two *Stylogaster* flies are copulating in the periphery of a *Labidus praedator* swarm raid. Henri Pittier National Park, Venezuela.

(Figures 3.33 and 3.34). *Stylogaster* eggs look like tiny harpoons, perfectly designed for being attached to host insects. However, which insects exactly are the hosts of *Stylogaster* is not entirely clear. It has been suggested that the eggs are being targeted at cockroaches and possibly other arthropods; at least in some cases, *Stylogaster* eggs have been found on swarm-attending tachinid flies (Rettenmeyer 1961; Rettenmeyer et al. 2011).

Arguably the most abundant and conspicuous swarm visitors among the Neotropical flies are species of the tachinid genus *Calodexia*. They often appear in the hundreds, perched on low vegetation and leaf litter in front of the advancing swarm front (Figure 3.35). The flies frequently shift position in order to follow the ants while simultaneously evading attack. Unlike in *Stylogaster,* almost all the *Calodexia* flies at an ant swarm are female. These females dart at fleeing cockroaches and crickets to deposit not eggs, but larvae—they are larviparous. The *Calodexia* larvae then burrow into their victim, where they feed voraciously on

FIGURE 3.35 A parasitoid fly, probably in the genus *Calodexia*, is perched over an *Eciton burchellii* swarm raid.

their unfortunate host while undergoing development (Rettenmeyer 1961; Rettenmeyer et al. 2011). It has been estimated that 50 to 90 percent of the cockroaches and crickets that manage to escape from the ants are being parasitized and ultimately killed by the flies instead (Rettenmeyer et al. 2011).

Most of the scuttle flies at army ant raids, on the other hand, are attracted by the alarm pheromones of the distressed prey ants, in which they deposit their eggs. Some even target the army ants themselves. This group of flies is highly diverse, and often quite specialized on particular

FIGURE 3.36 Several species of parasitoid scuttle flies (family Phoridae) attend army ant raids, and are frequently perched along raiding trails or hovering over the raid front. Although the hosts of most of these poorly studied insects remain unknown, many appear to parasitize the prey ants, rather than the army ants themselves (Brown and Feener 1998). Here, three scuttle flies, probably of the genus *Apocephalus*, are hovering above the entrance of a *Camponotus* carpenter ant nest that is under attack by *Eciton burchellii* (see Figure 3.15). The common name for *Apocephalus* is "ant-decapitating flies," owing to the fact that most species develop inside the heads of ants (e.g., Brown et al. 2018).

host ant species (Brown and Feener 1998; Disney et al. 2008). Scuttle flies often hover over the battlefield or perch along army ant raiding columns (Figure 3.36). Parasitoid scuttle flies have also sporadically been reported from army ant emigration columns (Disney et al. 2009a). Furthermore, several species of flesh flies specifically seek out *Eciton burchellii* swarm raids. Although their biology has not been studied in any detail, it has been suggested that they deposit their larvae on the occasional small frog or lizard that the ants kill but are unable to tear into pieces for transport (Rettenmeyer et al. 2011).

A few additional animals attend swarm raids in the Neotropics, at least occasionally and at some sites. For example, some parasitic wasps seem to seek out *Eciton burchellii* raids selectively, but exactly what they gain from the ants' proximity remains to be clarified (Rettenmeyer et al. 2011). Finally, in Brazil, several species of marmosets regularly feed on flushed insects at *Eciton burchellii* and *Labidus praedator* swarms (Rylands et al. 1989; Martins 2000; de Melo Júnior and Zara 2007).

Swarm raids of African driver ants are likewise attended by flies, including some *Stylogaster* species as well as blowflies (Calliphoridae) of the genus *Bengalia*. *Bengalia* flies are predators rather than parasitoids like *Stylogaster, Calodexia,* or scuttle flies, whose larvae develop inside a host insect. They are endowed with a particularly rigid, serrate proboscis that allows them to dart down at raids and emigrations to spear and steal prey and brood items from the ants, which they then consume (Gotwald 1995).

This dependence of many vertebrate and invertebrate species on swarm raids illustrates the role of army ants as keystone species in tropical rainforest ecosystems (Harper 1989; Brown and Feener 1998; Peters et al. 2008; Peters and Okalo 2009; Rettenmeyer et al. 2011). However, they might well exert their most severe ecological impact in their prominent role as arthropod predators.

The Ecological Impact on Prey Communities

As the predominant arthropod predators in tropical rainforests, it is safe to assume that army ants indeed play an important role in shaping the ecosystem (Rettenmeyer 1963a; Franks 1982a; Franks and Bossert 1983; Gotwald 1995; Boswell et al. 1998; Kronauer 2009, 2020; Rettenmeyer et al. 2011; Peters et al. 2013; Hoenle et al. 2019). But given how challenging it is to work with army ants experimentally, their precise effects on prey populations are difficult to quantify. One important factor to consider is the population density of army ant colonies. Unfortunately, for the vast majority of species we do not even have a vague idea of how common they are, given their elusive subterranean lifestyle. Nevertheless, it is clear that they can be extremely abundant. For example, the army ant *Labidus coecus* is the predominant ground-dwelling ant in some Neotropical rainforests (Ryder Wilkie et al. 2010). At least for the conspicuous swarm raider *Eciton burchellii,* several studies have measured population densities using a number of approaches, including standardized trail walking to encounter raids, as well as radio tracking ant-following birds. Using both methods at La Selva Biological Station, Vidal-Riggs and Chaves-Campos (2008) provided what is arguably the most reliable estimate of *Eciton burchellii* colony densities currently available: about six colonies per 100 hectares, or 247 acres. According to this study, you have a roughly 30 percent chance of encountering a raiding column while walking 1 kilometer of trail in the afternoon. Not bad, and certainly worth the effort.

Studies at other sites have generally reported slightly lower yet overall similar estimates: five colonies per 100 hectares in Cocha Cashu in Peru (Willson et al. 2011), and around 3.5 colonies per 100 hectares on Barro Colorado Island in Panama (Franks 1982a, 1982b). *Eciton hamatum* and *Eciton burchellii* occur at very similar densities, at least on Barro Colorado Island (Powell 2011). The population densities of African driver ants, on the other hand, can be considerably higher than those of *Eciton burchellii,* with estimates ranging from about twenty-five to fifty colonies per 100 hectares (Raignier and van Boven 1955; Leroux 1977b, 1982; Schöning et al. 2010). Accordingly, any given spot on the forest floor is raided about once every two months by *Dorylus molestus* at Mount Kenya (Schöning et al. 2010). Finally, the frequency of army ant raids can also vary on a finer spatial scale. At Kakamega Forest in Kenya, for example, the average chance for a local prey community to see a raid by one of the two resident driver ants, *Dorylus molestus* and *Dorylus wilverthi,* in any given week is just over 10 percent. However, this probability varied widely between sites during the study period, from 0 to 50 percent (Peters et al. 2013).

Another important factor is the number of different army ant species in a given ecosystem. Based on available inventories from La Selva Biological Station and Barro Colorado Island, a typical Neotropical rainforest site seems to be home to somewhere around twenty different army ant species, a remarkably high number indeed (Rettenmeyer 1963a; Longino et al. 2002; O'Donnell et al. 2007). Of course, the army ant species composition, and therefore the combined army ant raiding ecology and prey spectrum, can differ quite substantially between different sites (O'Donnell et al. 2007, 2009, 2011; Kaspari et al. 2011). However, summing over all observed species, army ant raids are not a rare occurrence at all (Kaspari and O'Donnell 2003; O'Donnell et al. 2007). For example, one study estimated that, averaged across four Neotropical rainforest sites, each square meter plot is hit by 0.73 army ant raids per day on average (O'Donnell et al. 2007). As we have seen, however, among the twenty or so army ant species at a given site, most will be fairly specialized in their diet, so the frequency of raids that affect any given prey species directly will be much lower (Powell and Franks 2006; Hoenle et al. 2019).

So what is the actual impact of army ant raids on prey populations? In a study of arthropod abundance in leaf-litter plots at La Selva Biological Station before and after *Eciton burchellii* raids, Gard Otis and colleagues indeed found that significantly fewer arthropods overall, as well as fewer different types of arthropods, were present after the ants had foraged. At the same time, the ants had seemingly overlooked a sizeable proportion of potential prey items. For example, the abundance of cockroaches was reduced by only 38 percent, quite similar to the 50 percent reduction found earlier by Nigel Franks on Barro Colorado Island (Franks 1982a; Otis et al. 1986). Furthermore, they found that about 80 percent of arthropods at the swarm front were able to escape; even among cherished prey items like crickets and cockroach nymphs, 42 and 36 percent were still able to make a run for it. A more recent study, however, found no consistent effect of *Eciton burchellii* swarms on invertebrate density or biomass (Kaspari et al. 2011).

Similarly, the effects on local arthropod communities of the subterranean Asian army ant *Dorylus laevigatus* are moderate to weak (Berghoff et al. 2003b), and so is the impact of the driver ant *Dorylus molestus* on the populations of its predominant prey, earthworms, at Mount Kenya (Schöning et al. 2010). *Labidus praedator* swarms, on the other hand, were found to leave behind a 25 percent reduction in total invertebrate biomass on average (Kaspari et al. 2011). Interestingly, the latter study also showed that the impact of *Labidus praedator* raids is not dependent on the initial prey density whereas the impact of *Eciton burchellii* is relatively higher in high-quality patches, suggesting that *Labidus praedator* is the more efficient predator at low prey abundance (Kaspari et al. 2011). This finding of relatively mild immediate impacts, together with the previously discussed insight that army ant raids are very common in tropical rainforest ecosystems (Kaspari and O'Donnell 2003; O'Donnell et al. 2007), led the authors to conclude that, rather than episodically "draining the bottle," army ants "skim the cream" by constituting a chronic but moderate, rather than an occasional yet catastrophic, type of predation pressure (Kaspari et al. 2011).

Mobile solitary arthropods quickly repopulate patches thinned out by army ants, and the precise long-term effects on those species are therefore difficult to measure; however, the sedentary colonies of social insects can be followed more easily over time. At La Selva Biological Station, for example, *Polistes* paper wasps frequently build their nests on the eaves of wooden buildings and are thus accessible for observation. In one such case, thirty-four active nests had been constructed by *Polistes erythrocephalus* in a line along one of the cabins, forming a happy population of larvae, pupae, and a total of 244 adults (Young 1979). Then the ants came. Like most other wasps, *Polistes* are unable to successfully defend their colonies against army ants, and they respond to the attack by evacuating the nest, leaving behind the helpless brood. The *Eciton burchellii* swarm attacked and plundered seventeen of the nests entirely, leaving not a single wasp larva or pupa behind alive. This attack also resulted in a decline of the adult wasp population, as many wasps probably deserted their nests to start over elsewhere. Three and a half weeks later the ants were back, and the wasp population declined further (Young 1979).

On Barro Colorado Island, Nigel Franks estimated that, following an *Eciton burchellii* raid, the prey ants require about 100 days to rebound to half their initial abundance (Franks 1982a). Combining this information with data on colony movements in a simulation model suggested that, at any given time, the prey ants on approximately half the island are recovering from army ant attacks (Franks and Bossert 1983). Clearly, *Eciton burchellii* raids can have quite detrimental effects on local social insect populations. The recovery time of prey populations might in turn feed back on colony abundance of a given species, in that army ants that predominantly consume solitary organisms reach higher population densities than those that mostly take social insects. This could help explain the striking differences in colony densities between *Eciton burchellii* and African driver ants, for example (Boswell et al. 2001; Franks 2001).

It has been suggested that the constant predation pressure from army ants in tropical ecosystems maintains social insect communities in a state of succession by disproportionally affecting and cropping dominant species. This is thought to regularly open up ecological opportunities for other species, thereby increasing the diversity of the local community (Franks 1982a; Franks and Bossert 1983; Kaspari and O'Donnell 2003; Powell and Baker 2008). Spatial variability in raid rates within a local ecosystem may help further increase the diversity of invertebrate communities globally by creating patchy patterns of species distributions and diversifying ecological niches locally (Peters et al. 2013). Finally, in addition to the previously discussed behavioral adaptations, it has been suggested that certain life-history traits of tropical leaf-litter ants can be interpreted as adaptations to army ant predation. For example, the colonies of tropical ants are on average smaller than those of temperate species, and their growth rate seems to be constant, rather than leveling off at large sizes. Both of these traits would be selected for under high and constant predation risk (Kaspari and O'Donnell 2003).

It has to be kept in mind, however, that army ant species composition and colony density, and thus their impact on local prey populations, can vary dramatically between different sites (O'Donnell and Kumar 2006; O'Donnell et al. 2007, 2009, 2011; Willson et al. 2011). It is therefore difficult to draw general conclusions. And even for the best-studied sites the data on the ecological impact of army ants remain rather anecdotal. The gold standard for understanding the ecological effects of army ants as top predators would involve long-term exclusion studies in which army ants are experimentally removed from replicate plots but not from control plots. Although such studies have been conducted for other predators (e.g., Ford and Goheen 2015), no rigorous attempts in this direction have been undertaken for army ants.

The essence of what makes an army ant is their collective predation in large numbers. But, as we have discussed in Chapter 2, the evolution of mass raiding required several additional adaptations to make this lifestyle sustainable. As we have seen, army ant raids can have an immense impact on the structure of tropical arthropod populations, and prey would likely quickly become exhausted if army ant colonies were restricted to hunt within the limited periphery of a stationary nest. However, army ant colonies are not subject to this constraint: they are nomadic, frequently relocating their bivouac nests to new feeding areas (Wheeler 1910; Wilson 1958a; Schneirla 1971; Franks and Fletcher 1983; Gotwald 1995; Kronauer 2009, 2020). Nomadism, the second pillar of the army ant adaptive syndrome, is the topic of the next chapter.

CHAPTER FOUR

Nomadism

Since dusk, I have been sitting on a small folding stool, my headlamp pointed at the emigration column of an *Eciton burchellii* colony passing by close to my rubber boots. At this point, I have watched intently for several hours, and when I close my eyes, the never-ending stream of ants simply continues as a dizzying hallucination. Tonight's emigration column traverses over a hundred meters across the forest floor, only to disappear into the thicket of a large treefall gap. A plume of skatole, an organic compound that is not only prominent in feces, but also gives army ants their characteristically sweet, musky odor, emanates from the dense undergrowth (Brown et al. 1979). I have given up on finding out exactly where the bivouac is located. Instead, I have perched myself a few meters before the green mass under which the ant highway vanishes. While early in the emigration the ants were mainly carrying prey, they are now exclusively transporting their own larvae; hundreds of thousands of them (Figure 4.1).

After the ants have gone about their business largely indifferent to my presence for hours, they are now getting more and more stirred up. So far, I have seen only a few soldiers patrolling along the trail, but at this point, there are several of them scouting the surroundings for intruders. As they venture farther and farther away from the emigration column, I am forced to back off a little. Suddenly, the regular traffic flow decreases markedly, and no more larvae are being carried. Instead, the trail is now swarming with soldiers, running back and forth in great excitement. And there she comes: covered in a protective mass of ants, the queen slowly makes her way down the path. Observing her walk freely in an emigra-

FIGURE 4.1 *Eciton burchellii* emigration columns resemble multilane highways. This picture shows the early stage of an emigration: while most workers are already carrying the colony's larvae, some are still transporting prey items.

tion is a great challenge because her entourage, consisting dispropor-tionally of soldiers, immediately forms a living shield around her at even the slightest disturbance (Figures 4.2 and 4.3). In fact, unless you are extremely careful not to alarm the ants, the workers will often sequester the queen for hours in some hideout along the trail. The queen and her entourage will then only complete the journey once all the other ants have long reached the new bivouac, and the disappointed human observer has abandoned his perch and gone to sleep. The army ants make sure that the most important member of their society travels safely. And the stakes are high indeed: only the queen can mate and lay eggs, and if she dies unexpectedly the entire colony will soon perish.

FIGURE 4.2 During the nomadic phase, *Eciton burchellii* queens are contracted and walk along in the nightly colony emigrations. Catching a glimpse of her majesty is challenging, however: she is frequently covered by a royal guard of soldiers and submajors. In the lower left corner of this image, the dark brown head of the queen is peeking out from under a mass of workers.

FIGURE 4.3 Army ant queens are heavily guarded when walking in emigration columns. This photograph shows an *Eciton hamatum* queen, all the way on the left, surrounded by her entourage, which disproportionally consists of soldiers. Large numbers of soldiers are also securing the trail in front of her.

Setting Up Camp

Many ant species periodically relocate their nests (McGlynn 2012). For example, many invasive ants nest rather opportunistically in exposed sites that are subject to frequent disturbance, and they will readily abandon their temporary home. Such ephemeral nesting habits probably contribute to their success as hitchhikers on human traffic around the globe (Tsutsui and Suarez 2003; McGlynn 2012). But even species that construct the most elaborate and valuable nests, such as leaf-cutting ants, can be incentivized to move under certain conditions, including partial destruction of the colony or other traumatic events (McGlynn 2012). Nevertheless, army ant colonies relocate much more frequently and over much larger distances than the colonies of other ants. In fact, nest relocations are an integral part of army ant biology.

Army ants are truly nomadic. However, unlike migratory birds and many other animals that seasonally travel between breeding and over-wintering grounds, army ants do not migrate, they emigrate (Heape 1931). Once an army ant colony has left an area, it is impossible to predict when, if ever, it will return. The nomadic lifestyle of army ants is reflected in the fact that their temporary bivouac nests can be rapidly disassembled, moved, and reassembled as needed. This practical property of the army ant nest was aptly put into context by Thomas Savage, a missionary and physician-naturalist who studied *Dorylus* driver ants in Africa (Savage 1847): "There is an old saying, which is not without meaning, that 'a man's dwelling indicates the nature of his employment.' A robber's house will not exhibit, either in or out, the indications of a permanent abode that an honest man's does; so with that of the insects before us, their mode of life will not admit of cells and magazines and other interior arrangements by which the domicils of other ants more retiring and less aggressive in their habits are characterized" (4). Although army ants do not construct sophisticated permanent nests like many other ants, we will see that their bivouacs, composed of many thousands or even millions of physically interlinked live ants, arguably constitute the epitome of William Morton Wheeler's famous analogy of the ant colony as an organism (Wheeler 1911). So how and when do bivouacs assemble and disassemble?

In the later afternoon, prey-laden *Eciton burchellii* workers are still returning home from the afternoon raid, but at this point the swarm activity has usually died down and hardly any workers are leaving the bivouac to join the foray. However, at around dusk, excitement inside the nest increases until workers once again start exiting the bivouac. As the amount of outbound traffic increases, so does the frequency of head-on collisions with homebound workers carrying booty. As we have seen in Chapter 3, more and more of the inbound workers therefore turn around, until the traffic flow is reversed and once again headed outward, typically along the base column of the afternoon raid. Initially, the outbound workers carry only booty from the day's raid. The early stages of this exodus can often be observed even though the colony will not ultimately move, and eventually all workers will return to the nest. However, if strong outbound movement is sustained, chiefly by sufficient stimulation from the developing army ant larvae, it will give rise to a colony emigration (Schneirla 1940, 1971).

At La Selva Biological Station, emigrations typically begin between 4:00 p.m. and 7:00 p.m., but especially if it rains heavily during this time window, onset can be delayed to as late as 10:00 p.m. (Califano and Chaves-Campos 2011). A reliable sign that an *Eciton burchellii* colony is in the process of relocating the nest is the appearance of workers carrying the colony's brood, typically larvae (see Figure 4.1). Ant brood also constitutes a prominent fraction of *Eciton burchellii* prey, but the army ants' own young can easily be recognized with a little bit of experience. Army ant larvae are long and slender, and in an emigration they occur in much larger numbers and are much more homogeneous in appearance than the brood of prey ants carried on raiding trails. As they do with prey items, army ants carry their larvae elegantly slung beneath their bodies, with the larval mouthparts facing forward and up.

At this point in an *Eciton burchellii* emigration, the column of ants has become a highway with approximately five lanes, all traveling in the same direction (see Figure 4.1). Only at the very periphery of the trail are scattered ants still running in the direction of the old bivouac. Soldiers now appear at strategic spots along the ant column to guard against unwelcome trespassers. Guards positioning themselves at particularly busy and exposed segments of emigration trails are a common phenomenon in all army ants (see Figure 2.16 and Figures 4.4–4.6). Any disturbance causes an immediate alarm response, with workers alerted by volatile mandibular gland pheromone swarming the surroundings of the column (Brown 1960; Keegans et al. 1993; Lalor and Hughes 2011; Brückner et al. 2018).

FIGURE 4.4 The sabre-shaped mandibles are the trademark of the iconic *Eciton* soldier. Here, an *Eciton burchellii* soldier is guarding the emigration of her colony.

FIGURE 4.5 Soldiers often pose along *Eciton* colony emigrations. Here, an *Eciton hamatum* soldier seems to oversee the orderly conduct of the procedure, with regular workers carrying larvae at the bottom of the image in typical army ant-fashion: slung beneath the body, with the larval mouthparts facing the worker's head.

FIGURE 4.6 A *Dorylus molestus* soldier is standing guard at the periphery of an emigration column at Mount Kenya, Kenya.

Because traffic flow is usually highest during colony emigrations, this is also when the most impressive army ant suspension bridges emerge (Figure 4.7). Where the ant column traverses steep terrain or thin branches, workers position themselves to form elaborate flanges to keep their laden sisters from slipping, another form of living architecture discussed in Chapter 3 (Figure 4.8).

FIGURE 4.7 Army ant traffic is particularly heavy during colony emigrations, resulting in impressive examples of bridge formation, as discussed in Chapter 3. This *Eciton burchellii* suspension bridge not only allows greater traffic flow by widening the path for the workers carrying larvae, but it provides a shortcut over two twigs intersecting at a right angle.

FIGURE 4.8 When heavy traffic traverses slippery terrain, *Eciton burchellii* army ants form support structures in the form of living flanges. Here, the ants are employing this trick to expand a narrow branch over which an emigration is moving. The same phenomenon occurs when the ants move over steep surfaces such as tree trunks, which can be seen in the foreground.

Localized brawls may arise wherever the nightly path through the rainforest conflicts with the emigration columns of other army ants or the nocturnal harvesting activity of leaf-cutting ants (Figures 4.9–4.12). Some of the booty caches described in the previous chapter are now turning into brood caches along the emigration column as the ants deposit more and more larvae. These roadside restaurants, where larvae often snack on prey items, are once again heavily guarded by soldiers (Figures 4.13 and 4.14).

FIGURE 4.9 The tropical rainforest provides sufficient structure to allow army ant traffic in three dimensions. This comes in handy when the paths of two colonies meet. Here, an *Eciton burchellii* emigration (on the right) is bridging an orthogonal *Neivamyrmex gibbatus* emigration (on the left) by traveling over a leaf on the forest floor. Yet this image also reveals several other interesting features of army ant societies. Young army ants in phasic species eclose in discrete cohorts. Initially, these callow workers are lighter in color than their older sisters, because their cuticle is not fully melanized and hardened yet. This is illustrated here by the bright yellow callow *Neivamyrmex gibbatus* workers walking in the emigration. Furthermore, the image shows a fundamental organizational principle of insect societies: age polyethism, the division of labor as a function of age. Most of the ants walking in the *Neivamyrmex gibbatus* emigration are callows, and only their older and darker sisters are guarding the emigration against *Eciton burchellii* workers. Similarly, army ant callow workers do not participate in raids, and in the clonal raider ant *Ooceraea biroi*, age correlates with an ant's tendency to leave the nest (Topoff et al. 1972b; Topoff and Mirenda 1978; Ulrich et al. 2020). This phenomenon of age-related differences in behavior is well understood in honeybees and some other ants, but remains relatively poorly studied in army ants. Finally, a staphylinid beetle in the genus *Tetradonia* is walking along the *Neivamyrmex gibbatus* emigration in the background. These myrmecophiles, or ant guests, will be the subject of Chapter 6.

FIGURE 4.10 The course of nightly army ant highways sometimes conflicts with the movements of other colonies. This can result in confined local confrontations that usually do not escalate. This picture shows a standoff between the army ants *Eciton burchellii* (left) and the smaller *Neivamyrmex gibbatus* (right) around the area where the emigration columns of two colonies collide. This image is a close-up of the scene depicted in Figure 4.9.

FIGURE 4.11 At the intersection of two army ant emigration columns, an *Eciton hamatum* worker (right) is being held at bay by a group of black *Neivamyrmex pilosus* workers (left).

FIGURE 4.12 Army ant traffic can also collide with the foraging paths of other dominant social insects. Here, an emigration column of *Eciton burchellii* blocks the harvesting activities of *Atta cephalotes* leaf-cutting ants. These encounters do not escalate but can result in substantial traffic jams and rerouting.

FIGURE 4.13 A brood cache has formed along an *Eciton burchellii* emigration. The actual emigration column is visible in the background.

FIGURE 4.14 Army ant brood caches are often guarded by soldiers, as is the case here for *Eciton hamatum.*

From around 7:30 p.m. onward, some of the more distant ant clusters are growing in size until they form nascent bivouacs (Schneirla 1971). These early stages of bivouac construction usually emerge at a few sites simultaneously. The ants seem to evaluate various options before the colony collectively decides where to move. Only later in the emigration will it become apparent which of these clusters will be the new bivouac, and sometimes the colony decides to continue its exodus even after most of the ants have settled at a given site. Because army ants are difficult to manipulate experimentally, and it is not trivial to keep track of which sites exactly are being scoped out by the ants at any given point in time, very little is known about this collective decision-making process. We do know that it occurs remarkably regularly and predictably, and constitutes an integral part of army ant life history. Furthermore, microclimatic considerations seem to be important (Soare et al. 2011), although it is unclear which properties of potential nesting sites exactly are relevant to the ants' decision making, let alone how the ants weigh different attributes and communicate and integrate the information on the quality of available sites.

These fascinating questions have been studied in considerable detail during swarming in honeybees and house hunting in *Temnothorax* ants, which constitute among the best understood examples of distributed decision making (e.g., Franks et al. 2002; Visscher 2007; Seeley 2010; Sasaki and Pratt 2018). Generally, scouts find and evaluate the quality of different possible nesting sites. Their assessment is then communicated to nestmates. The level of excitement for a given site is usually encoded in the amount of vigor with which they recruit additional scouts. The alternative nest sites then compete for scouts, and once one site has amassed a sufficiently large and disproportionate number, the colony reaches a quorum and moves.

Eciton burchellii colonies often construct their bivouacs in relatively exposed locations at ground level, but always against some structure that provides a scaffold and protection from the heavy downpours that are common in tropical rainforests. Typically, these are fallen tree trunks, open cavities in the base of trees, or the buttress roots of tree giants (Schneirla 1971; Teles da Silva, 1977b; Kronauer et al. 2007b) (see Figure P.5 and Figure 4.15). Neither are the ants opposed to man-made media: I have repeatedly seen colonies camp out in wooden or concrete sheds. *Eciton hamatum* uses similar nest sites but can also construct bivouacs on the forest floor using entirely leafy vegetation as scaffolding (Figure 4.16).

FIGURE 4.15 Free-hanging army ant bivouacs are cone-shaped. Here, an *Eciton burchellii* bivouac is suspended from the underside of a fallen tree trunk.

FIGURE 4.16 When the ants reach the ground during bivouac formation, the resulting shape is generally cylindrical but in its details dictated by properties of the substrate. Here, an *Eciton hamatum* colony has assembled its bivouac under a branch and two large adjacent leaves of the same plant, giving rise to a particularly striking architecture that resembles a cathedral with three naves.

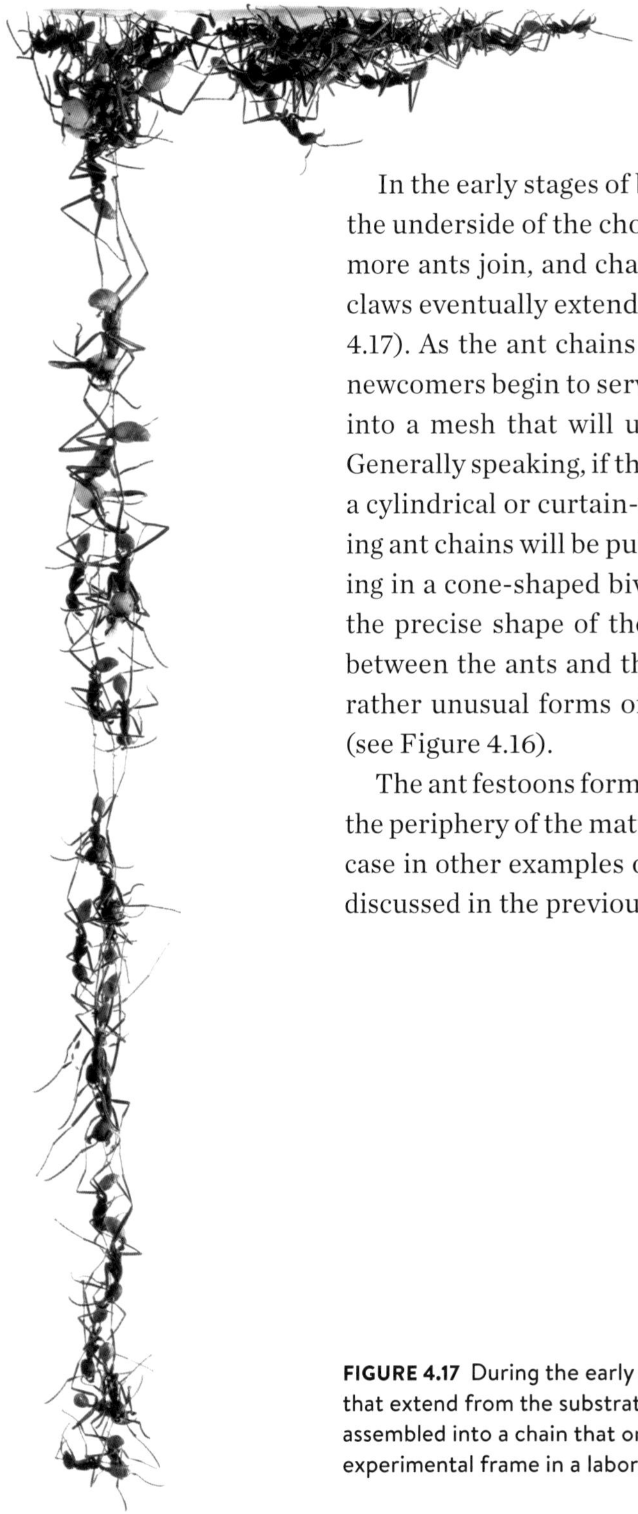

In the early stages of bivouac construction, the ants form clusters on the underside of the chosen substrate (Schneirla 1933, 1971). More and more ants join, and chains of ants that are interlocked via their tarsal claws eventually extend from these clusters toward the ground (Figure 4.17). As the ant chains grow into ropes and become more numerous, newcomers begin to serve as crosslinks, slowly connecting the festoons into a mesh that will ultimately form the outer wall of the bivouac. Generally speaking, if the chains reach the floor below, the result will be a cylindrical or curtain-style bivouac (see Figure P.5). If not, the growing ant chains will be pulled inward as they are being connected, resulting in a cone-shaped bivouac (see Figure 4.15). However, in each case, the precise shape of the final bivouac emerges from the interactions between the ants and the substrate, which sometimes can give rise to rather unusual forms of army ant architecture (Schneirla et al. 1954) (see Figure 4.16).

The ant festoons forming the skeleton of the bivouac are still visible at the periphery of the mature installation (Figures 4.18 and 4.19). As is the case in other examples of living army ant architecture, like the bridges discussed in the previous chapter, the tension exerted by the structure,

FIGURE 4.17 During the early stages of bivouac construction, army ants form chains that extend from the substrate downward. Here, *Eciton burchellii* workers have assembled into a chain that originates from a cluster of ants on the underside of an experimental frame in a laboratory setup.

possibly in combination with traffic flow of mobile ants, maintains the participating workers' immobile engagement. *Eciton* army ants can be induced to form bivouac-like structures in the laboratory. When housed in narrow frames, the resulting construction gives the impression of a bivouac cross section, providing a unique insight into the inner workings of the nest, where hollow chambers packed with brood are separated by

FIGURE 4.18 Isolated ant chains are still visible at the periphery of bivouacs, here in a close-up of the *Eciton burchellii* bivouac shown in Figure 4.15.

FIGURE 4.19 Several isolated ant festoons also help attach the *Eciton hamatum* bivouac shown in Figure 4.16 to its leafy substrate.

FIGURE 4.20 When placed inside an experimental observation frame, army ants assemble into structures that resemble thin bivouac cross sections. Brood chambers, passageways, and ant meshes appear in this photograph of an *Eciton hamatum* colony.

passageways and galleries lined by the bodies of interwoven workers (Schneirla 1971) (Figures 4.20 and 4.21).

The bivouac provides an effective shelter against rain and predators for the queen and her offspring. If an unsuspecting human breathes against the structure, the surface springs into action and trembles as if to warn the putative attacker (Beebe 1919). But the bivouac also constitutes an incubator for the brood. Accordingly, *Eciton burchellii* bivouacs are tightly climate controlled, with only relatively small temperature and humidity fluctuations compared with considerably larger variations in ambient parameters. To maintain favorable conditions throughout the day, the ants loosen or tighten the live fabric of the bivouac and

FIGURE 4.21 A close-up of the observation frame from Figure 4.20 shows two brood chambers separated by a passageway.

open and close ventilation vaults, allowing more or less of their metabolic heat to dissipate and water to evaporate (Schneirla et al. 1954; Jackson 1957; Teles da Silva 1977b; Franks 1989b; Baudier and O'Donnell 2016; Baudier et al. 2019). This capacity to thermoregulate is probably also important for the well-being of the adult ants themselves, which must ensure that temperatures do not exceed their thermal tolerance (Meisel 2006; Baudier and O'Donnell 2018; Baudier et al. 2018).

As the new bivouac emerges, the old one disassembles. With the ongoing exodus of workers, it slowly becomes hollow, until only the outside sheets remain. Eventually, the remaining walls of ants become detached from the bottom substrate and melt away upward. The living walls can also collapse in the process of dismantling the bivouac, resulting in large clusters of ants and brood at the base of the emigration column (Schneirla 1971). The queen and her entourage typically leave the old bivouac only after more than half of the brood has been sent on its way, usually not before 8:00 p.m. (Schneirla 1971).

Eciton burchellii emigrations often last between six and eight hours, but the precise duration depends on a lot of factors, such as the colony size, the distance covered, and whether the process gets disrupted by adverse weather conditions (Schneirla 1971; Califano and Chaves-Campos 2011). While some emigrations are completed by midnight, others last until the early morning hours. In the vast majority of cases, however, the colony has concluded its relocation before dawn, and at daybreak, no visible signs of the nightly procession remain.

Reproduction and Brood Development in Phasic Species

Army ants have been portrayed as being "here today and gone tomorrow," but this is not entirely accurate, at least for some species. It turns out that the emigrations of *Eciton burchellii, Eciton hamatum,* and many other army ants are in a sense highly predictable. Colonies of these species undergo stereotypical cycles in which they alternate between so-called nomadic and statary (an old English term meaning "settled") phases. The colony cycles have two main components. The first arises from cyclic changes in the reproductive physiology of the army ant queen. She lays eggs only during the statary phase, and the resulting larvae develop during the nomadic phase. The second component concerns cyclic changes in the nomadic behavior of the colony, which parallel the reproductive cycle of the queen. During the statary phase, the colony stays put, while the ants emigrate to a new bivouac site during most nights of the nomadic phase. The statary and nomadic phases of *Eciton burchellii* colonies last for about three and two weeks, respectively.

Studying the army ants' nomadic behavior is not without its challenges. Not unlike scientists tracking troops of primates through the rainforest, army ant researchers have to be constantly on the move. But unlike large mammals, army ant colonies cannot be radio-tagged, and once a colony has been lost, even momentarily, it is usually impossible to find it again and ascertain its identity. The colony cycles of phasic army ants therefore remained elusive well into the twentieth century.

Wilhelm Müller, who had discovered the first *Eciton burchellii* male inside a colony in his brother's garden in Blumenau, was also the first to come close to discovering the cyclic behavior of this species (Müller 1886). Müller painstakingly followed his focal colony for eighteen days, approximately half the length of a typical colony cycle, and he witnessed the transition from the nomadic to the statary phase. Crucially, Müller repeatedly collected brood from the colony and kept track of the occurrence of eggs, larvae, and pupae. He noticed that, as more and more larvae pupated, the colony stopped emigrating, and the intensity of the daily swarm raids decreased. He also noticed that, shortly after the emigrations had ended, eggs appeared in the colony. He thus concluded that a pregnant female must have been present in the colony at that point. Based on the immense size of army ant colonies, he went on to surmise that this female was probably very large and immobile in her egg-laying state. Unfortunately, Müller was forced to abandon his studies toward the end of the statary phase of his colony, and he was therefore not able to observe a colony cycle in its entirety.

In the summer of 1932, Theodore Schneirla, a comparative psychologist at New York University who would later also become curator at the American Museum of Natural History, traveled to Barro Colorado Island in Panama to study army ants. The two months he spent that summer roaming the rainforest (Schneirla 1933) were the prelude to an unprecedented career in army ant research that spanned more than three decades and culminated in Schneirla's posthumously published book *Army Ants: A Study in Social Organization* in 1971. Much of what we know today about the biology of *Eciton burchellii* and *Eciton hamatum* is based on Schneirla's work. It was during his second field season in Panama in 1936 that Schneirla became the first researcher to track a single colony of *Eciton hamatum* through an entire colony cycle (Schneirla 1938). He later demonstrated that several other army ants, including *Eciton burchellii,* show the same cyclic behavior (e.g., Schneirla 1944a, 1944b, 1945, 1947, 1957a, 1957b, 1963, 1971; Schneirla and Brown 1950; Schneirla and Reyes 1966, 1969) (see Table 2.3).

To understand army ant colony cycles we must recall insect development. As a brief primer, Hymenoptera, along with flies, beetles, and butterflies, are holometabolous insects. Holometabolous insects undergo complete metamorphosis with distinct larval and pupal stages. The larva hatches from the egg, and as it feeds and grows it can undergo several molts that subdivide larval development into discrete larval instars, or developmental stages. The larval stage ends, and the pupal stage begins, when the larva stops feeding and pupates. The pupa can either be naked or enclosed in a cocoon, depending on the species. Even among ants there is variation in this respect. In African *Dorylus* driver ants, for example, pupae are naked, while *Eciton* pupae are enclosed in silken cocoons spun by the larvae. But whether or not they are enclosed in cocoons, the pupae do not feed. Instead, it is during the pupal stage that the organism becomes entirely remodeled. Eventually, the adult, also called the imago, emerges from the pupa. The technical term for this step is eclosion. In ants, the newly eclosed adults are called callows, and are usually lighter in color because their cuticle—the outer covering of an insect and its exoskeleton—has not completely hardened yet (see Figure 4.9).

In *Eciton burchellii* and *Eciton hamatum,* larvae that eventually become workers undergo five larval instars that can be distinguished by size and the presence and distribution of hairs on different parts of the body (Wheeler and Wheeler 1986). Approximately two-thirds into the nomadic phase, an *Eciton burchellii* colony contains larvae ranging from about the second to the fifth larval instar (von Beeren et al. 2016b). Especially the fifth instar larvae already display quite a substantial range in body length, from 3.4 to 12.0 millimeters, foreshadowing their level of polymorphism as adults (Schneirla et al. 1968) (Figure 4.22). Despite the

FIGURE 4.22 While the level of brood synchronization in phasic army ants is remarkable, it is not perfect. This picture shows the range of developmental stages collected from an *Eciton burchellii* emigration during the mid- to late nomadic phase. From left to right, these are second, third, fourth, and two fifth instar worker larvae. The two fifth instar larvae illustrate the substantial size polymorphism present at this final stage of larval development. The size at the onset of pupation determines the caste fate of the adult. In other words, the smallest larvae will become small workers, while the largest larvae will become soldiers. Detailed descriptions of army ant larval morphology and development are provided by Wheeler (1943), Tafuri (1955), Lappano (1958), Wheeler and Wheeler (1964, 1974, 1984, 1986), and Schneirla et al. (1968).

fact that different larval instars co-occur during the nomadic phase, the broods are remarkably synchronized in their development. While in most nonseasonal ant species all developmental stages are found simultaneously, eggs and pupae are entirely absent from a few days into the *Eciton burchellii* nomadic phase onward. The larvae feed voraciously on the food collected by the workers. The workers initially macerate and knead the prey into pellets, which they place on the larvae. As the larvae grow, the workers also put them directly on larger prey items, where they attach with their mouthparts (Schneirla 1971). The larval mandibles are incapable of chewing solid food, but instead it seems likely that the larvae exude enzymes to digest the prey externally and then suck up the resulting liquid (Wheeler and Wheeler 1984).

The nomadic phase ends, and the statary phase begins, when the larvae stop feeding and enter pupation. During the first days of the statary phase, workers bring larvae out of the bivouac to spin their cocoons. A colorful description of this process was provided by William Beebe based on his observations of an *Eciton burchellii* colony bivouacked at the Kartabo field station in British Guiana (Beebe 1919). The workers were placing the final instar larvae on an old wooden board in the station's storage room.

> On the flat board were several thousand ants and a dozen or more groups of full-grown larvae. Workers of all sizes were searching everywhere for some covering for the tender immature creatures. They had chewed up all available loose splinters of wood, and near the rotten, termite-eaten ends, the sound of dozens of jaws gnawing all at once was plainly audible. This unaccustomed, unmilitary labor produced a quantity of fine sawdust which was sprinkled over the larvae. I had made a partition of a bit of a British officer's tent which I had used in India and China, made of several layers of colored canvas and cloth. The ants found a loose end of this, teased it out, and unraveled it, so that all the larvae nearby were blanketed with a gray parti-colored covering of fuzz.
>
> All this strange work was hurried and carried on under great excitement. The scores of big soldiers on guard appeared rather ill at ease, as if they had wandered by mistake into the wrong department. They sauntered about, bumped into larvae, turned and fled. A constant stream of workers from the nest brought hundreds more larvae, and no sooner had they been planted and débris of sorts sifted over them, than they began spinning. A few had already swathed themselves in cocoons—exceedingly thin coverings of pinkish silk. As this took place out of the nest, in the jungle they

must be covered with wood and leaves. The vital necessity of this was not apparent, for none of this débris was incorporated into the silk of the cocoons, which were clean and homogeneous. Yet the hundreds of ants gnawed and tore and labored to gather this little dust, as if their very lives depended upon it. . . .

When first brought from the nest, the larvae lay quite straight and still, but almost at once they bent far over in the spinning position. Then some officious worker would come along, and the unfortunate larva would be snatched up, carried off, and jammed down in some neighboring empty space, like a bolt of cloth rearranged upon a shelf. Then another ant would approach, antenna the larva, disapprove, and again shift its position. It was a real survival of the lucky, as to who should avoid being exhausted by kindness and over-solicitude. . . .

There was no order of packing. The larvae were fitted together anyway, and meagerly covered with dust of wood and shreds of cloth. One big tissue of wood nearly an inch square was too great a temptation to be left alone, and during the course of my observation it covered in turn almost every group of larvae in sight, ending by being accidentally shunted over the edge and killing a worker near the kitchen middens. There was only a single layer of larvae; in no case were they piled up, and when the platform became crowded, a new column was formed and hundreds taken outside. To the casual eye there was no difference between these legionaries and a column bringing in booty of insects, eggs, and pupae; yet here all was solicitude, never a bite too severe, or a blunder of undue force. (463–464)

The temporary cover of debris observed by Beebe in fact plays an important role in the spinning of the pupal cocoons because it provides the legless ant larvae with a substrate to which they can attach their silk (Wheeler 1921).

As the larvae enter the pupal stage and the colony becomes statary, the *Eciton* queen embarks on a remarkable transformation. During the nomadic phase, her gaster is fully contracted, and she does not lay eggs (see Figures 1.10, 1.11, 4.2, and 4.3). Now, however, her gaster begins to swell up, and the intersegmental membranes between the sclerites become visible as thick white stripes. The queen is activating her ovaries. The technical term for this condition is physogastry, which refers to the unusual level of expansion of the queen's gaster. As Wilhelm Müller had predicted, the reproductively active *Eciton burchellii* queen is massive. Schneirla (1971) reported that the gaster of the same queen doubles in length, from 10.7 millimeters during the nomadic phase to 21.5 millimeters during the statary phase. At about a week into the statary phase, the queen has become fully gravid and begins to lay eggs (Figures 4.23 and 4.24). Over the next ten days or so, she lays the eggs that will give rise to the next cohort of larvae and eventually callow workers, about 300,000 in a typical *Eciton burchellii* worker brood (Schneirla 1971).

In her most expanded state, the queen is pretty much immobile, and even the workers are unable to move her over larger distances. The reproductive output of the *Eciton hamatum* queen is still impressive, yet somewhat smaller, with worker broods numbering around 80,000 individuals (Schneirla 1971). The development from egg to adult takes approximately forty-four days in *Eciton burchellii,* and forty-seven days in *Eciton hamatum* (Teles da Silva 1977a). Toward the end of the statary phase, the eggs begin to hatch, and the queen's gaster once again contracts. At around the same time, a cohort of callow workers emerge from the pupae, and the empty cocoons glide to the ground like miniature handkerchiefs, covering the colony's waste dump on the floor beneath the bivouac with a thick layer of golden silk (Figure 4.25).

As these events unfold, the ants become nomadic once again. In *Eciton burchellii,* the colony still contains many pupae during the first nomadic days, along with the eggs and early instar larvae of the next worker cohort. Interestingly, *Eciton* larvae undergo some sort of "developmental convergence," which further tightens their level of synchrony. The *Eciton hamatum* queen, for example, lays eggs over the course of about a week, but all larvae pupate and callows eclose within only two to three

FIGURE 4.23 During the statary phase, the gaster of the *Eciton burchellii* queen swells up to a remarkable extent, and her ovaries are now filled with eggs, making her daughter workers appear even tinier in comparison. The queen has become physogastric and is rather immobile in this condition. Because physogastric army ant queens do not leave the bivouac, this picture was taken in a laboratory setup after the queen and a small group of workers had been isolated from their colony.

FIGURE 4.24 Eggs emerge from the rear end of the physogastric *Eciton burchellii* queen in a continuous chain. Inside the colony, the smallest workers immediately pick up and care for the eggs. This picture was taken in a laboratory setup.

FIGURE 4.25 *Eciton burchellii* workers eclose from their cocoons toward the end of the statary phase and during the first few days of the nomadic phase. This picture shows a thick layer of empty pupal cases that has formed on top of the garbage heap beneath a statary bivouac inside a large tree cavity.

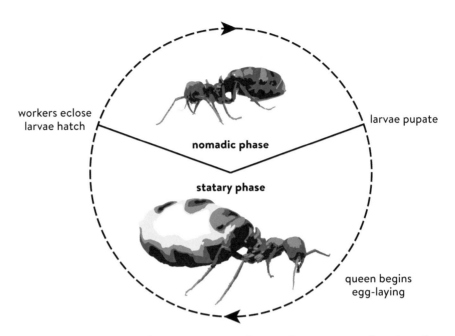

workers eclose
larvae hatch

nomadic phase

larvae pupate

statary phase

queen begins
egg-laying

FIGURE 4.26 The colonies of phasic army ants alternate between nomadic or brood care phases, and statary or reproductive phases. During the nomadic phase, the colony contains a cohort of developmentally synchronized larvae, the queen's gaster is contracted, and she does not lay eggs. When the larvae pupate, the colony enters the statary phase. During the statary phase, the queen's gaster expands until she is fully physogastric and lays a large batch of eggs. As the larvae hatch from the eggs, the queen's abdomen contracts once again, and the colony enters the next nomadic phase. This transition typically coincides with the eclosion of callow workers from the previous brood cohort. These dynamics are depicted here based on data from *Eciton burchellii* that are discussed in detail in the main text of this chapter.

days of one another (Schneirla 1949, 1971; Schneirla and Brown 1950; Schneirla et al. 1968). Although the *Eciton burchellii* queen in the first colony emigration looks nothing like she did a week earlier, her gaster can initially still be somewhat distended, with the intersegmental membranes partly visible (Schneirla 1971). However, at this point she is able to walk independently. Within a few days, all of the callow workers have eclosed, the queen's gaster has contracted completely, and the larvae are growing as they gobble the food retrieved and presented to them by their adult sisters. The colony is on its way to completing yet another reproductive cycle (Figure 4.26).

Colony Behavior in Phasic Species

Among the most arduous studies of army ant cyclic behavior must be that of Madalena Teles da Silva, who conducted her graduate research following colonies in the Amazonian forest of Belém (Teles da Silva 1977a, 1977b, 1982). Over a two-year period, Teles da Silva took notes on 137 *Eciton burchellii* and forty *Eciton hamatum* bivouacs. This involved following one particular *Eciton burchellii* colony continuously over nineteen complete colony cycles, from June 1966 until March 1968. Based on the thirty-five cycles observed across different *Eciton burchellii* colonies, Teles da Silva estimated an average cycle length of thirty-five days, with a range from thirty-one to forty-two days. Colonies spent fourteen days on average in the nomadic phase, with a range from ten to twenty days, and twenty-one days in the statary phase, with a range from sixteen to twenty-seven days. These estimates are in tight accordance with Schneirla's previous observations on Barro Colorado Island (Teles da Silva 1977a).

In *Eciton hamatum,* the colony cycles are similar in length, but considerably less variable than in *Eciton burchellii.* Based on the extensive data on *Eciton hamatum* collected by Schneirla on Barro Colorado Island, the average length of a nomadic phase is seventeen days, with a narrow range from sixteen to eighteen days among twenty-eight observed nomadic phases. The average statary phase lasts for twenty days, ranging from eighteen to twenty-two days based on thirty-nine records (Teles da Silva 1977a). The average length of the two phases can differ substantially between populations of the same species, however, as becomes apparent from studies of the North American *Neivamyrmex nigrescens* (Schneirla 1958; Mirenda and Topoff 1980). These discrepancies might be attributable to climatic differences and hence differences in developmental timing (Mirenda and Topoff 1980).

The extensive data collected by Schneirla, Teles da Silva, and others also reveal more subtle changes in behavior throughout the *Eciton burchellii* colony cycle. The first thing to note is that the sites at which colonies decide to construct their bivouacs change dramatically between the nomadic and the statary phase. Nomadic bivouacs are often exposed and openly suspended from structures such as logs or the buttress roots of trees, and the vast majority are located at ground level. Statary bivouacs, on the other hand, are typically constructed in more sheltered spaces such as large tree cavities and tend to be more elevated, with about two-thirds of them at a height of more than 1 meter (Teles da Silva 1977b). In fact, the majority of statary bivouacs I have personally observed were situated inside large, hollow trees, several meters off the ground and entirely inaccessible to the human observer. These differences may well reflect a trade-off between the stability and thermoregulatory properties of nesting sites, and their availability.

In the statary phase, the colony will occupy a given nest site for an extended period of time, and once the queen has become physogastric and is laying eggs, the nest cannot be easily evacuated in case of an emergency, such as physical disturbance or adverse climatic conditions. In the nomadic phase, on the other hand, the colony rarely remains at a site for more than one night, and can respond to perturbations by immediately relocating the nest. It can therefore afford to be less choosy in its house-hunting decisions. In fact, nomadic bivouacs tend to be less exposed during the dry season under more adverse climatic conditions, and it has been suggested that more sheltered bivouacs are better at maintaining a constant microclimate than more exposed ones (Teles da Silva 1977a).

In the North American army ant *Neivamyrmex nigrescens,* behavioral changes are also apparent in small groups of workers in the laboratory. Ants are more light-averse and have a higher tendency to cluster together in the statary phase, which might reflect some of the behavioral differences observed in the field (Topoff 1975).

Another aspect of colony behavior that changes quite dramatically between the two phases of the cycle is the frequency and intensity of the swarm raids. Raids occur essentially daily during the nomadic phase, but Teles da Silva (1982) found that *Eciton burchellii* colonies in Belém do not raid at all on 27 percent of the days during the statary phase. Raids during the nomadic phase are also more extensive and persistent than during the statary phase, at least in some populations. According to Teles da Silva (1982), the maximum length of the raiding column was 76 meters on average during the nomadic phase, but only 57 meters during the statary phase. A study in the Peruvian Amazon estimated that the average swarm raid during the nomadic phase is 12.1 meters wide, and the average width during the statary phase, while still impressive, is only 9.1 meters (Willson et al. 2011). Finally, colonies in the nomadic phase almost always raid during the entire day (98.5 percent of the cases; Teles da Silva 1982), but the raids of statary colonies sometimes last only for half a day (16 percent of the cases; Teles da Silva 1982). Overall similar patterns have been reported for *Eciton hamatum* (Teles da Silva 1982). Clearly, the army not only becomes settled, but also somewhat more docile during the statary phase.

Interesting changes in behavior also occur over the course of the nomadic phase. In *Eciton burchellii,* the first nomadic bivouac is often rather atypical in that it is sheltered, similar to statary bivouacs. This is where the majority of callow workers eclose from the pupae, and the colony often remains at the first nomadic bivouac site for several days (Teles da Silva 1977b). From now on, the colony will emigrate almost every night for the rest of the nomadic phase, although *Eciton burchellii* colonies occasionally still skip an emigration. The second nomadic bivouac is typically exposed. From here on the frequency of sheltered bivouacs increases throughout the nomadic phase, until the late nomadic bivouacs are sheltered once again, similar to statary bivouacs (Teles da Silva 1977b).

The changes in the level of bivouac exposure are paralleled by systematic changes in the lengths of emigrations throughout the nomadic phase. *Eciton burchellii* colonies emigrate in the direction of the day's raid, and the distance between two consecutive bivouacs is equal to or less than the distance covered by the raiding column, typically about 70 percent of the maximum raid length (Schneirla 1944b, 1945). In Teles da Silva's study in Belém, the average distance between consecutive bivouacs was 52 meters, and individual emigrations ranged from 10 to 180 meters in length (Teles da Silva 1977b). Emigrations at Belém seem to be somewhat shorter on average than in other populations, such as the Peruvian Amazon with 78 meters on average (Willson et al. 2011), or Barro Colorado Island with 81 meters on average (Willis 1967). The emigration that leaves the statary bivouac—that is, the first emigration of the nomadic phase—is usually shorter than the following emigrations, only around 40 meters on average (Teles da Silva 1977b).

Toward the end of the nomadic phase, on the other hand, the length of raids and emigrations increases markedly, and colonies cover approximately 120 meters on average in their final emigration to the next statary bivouac (Teles da Silva 1977b). Although it remains unclear whether this increase in raiding and emigration distance has any adaptive significance, it is intriguing to speculate that colonies develop longer raids when their larvae are larger and might require more food. Additionally, this pattern implies that colonies tend to establish their statary bivouacs far away from previous hunting grounds. After all, they will have to hunt sustainably around the statary bivouac site for the next three weeks.

However, some of the more subtle aspects of *Eciton burchellii* biology might differ between different populations, either due to environmental or genetic differences. For example, colonies in the Peruvian Amazon are much more likely to skip an emigration during the nomadic phase than colonies on Barro Colorado Island: colonies did not emigrate on 31 percent of nomadic days in Peru (Willson et al. 2011), whereas colonies on Barro Colorado Island failed to emigrate on only 14 percent of nomadic days (Willis 1967). Furthermore, unlike in Belém, colonies on Barro Colorado Island are not particularly likely to skip emigrations early in the nomadic phase (Willis 1967; Franks and Fletcher 1983). Finally, colonies of *Neivamyrmex nigrescens* are not more likely to skip emigrations early into the nomadic phase either, and emigrations do not systematically increase in length (Mirenda and Topoff 1980; Topoff 1984). It has therefore been suggested that the local availability of prey or nest sites might overshadow the putative effect of larval development on colony behavior (Mirenda and Topoff 1980; Topoff 1984).

Interestingly, both the raid and emigration columns of *Eciton hamatum* cover substantially larger distances than those of *Eciton burchellii*. The mean distance between consecutive bivouacs at the Belém site was 157 meters, with a range from 10 to 400 meters; as in *Eciton burchellii,* the distances of emigrations increased over the course of the nomadic phase (Teles da Silva 1977b). Similar to *Eciton burchellii,* the bivouacs tend to be more exposed in the nomadic phase than in the statary phase. In fact, *Eciton hamatum* colonies sometimes construct their nomadic nests simply under large leaves, forming what are arguably the most exposed army ant bivouacs of all (see Figure 4.16). However, unlike in *Eciton burchellii,* all callow workers emerge inside the *Eciton hamatum* statary bivouac, and colonies hardly ever skip an emigration during the nomadic phase (Schneirla and Brown 1950; Teles da Silva 1977b). It has been suggested that these differences in nomadic behavior, just like the differences in foraging behavior discussed in Chapter 3, are related to the fact that the diet of *Eciton hamatum* is much more specialized than that of *Eciton burchellii* (Teles da Silva 1977b). According to this idea, frequent emigrations that cover larger distances might help maximize the chance of encountering fresh colonies of the few wasp and ant species that make up the *Eciton hamatum* menu.

Another, somewhat contentious issue is whether raids and emigrations are random in direction (Franks and Fletcher 1983; Britton et al. 1996; Franks 2001; Willson et al. 2011; Califano and Chaves-Campos 2011). Intuitively, it would make sense for colonies to avoid overlap in the area covered by raids on consecutive days as much as possible. After all, there is little use in plundering a village that has just been sacked. In the statary phase, where the raids form spokes around the central hub of the bivouac, the problem becomes analogous to a plant growing new leaves from its vertical stem in a way that minimizes self-shading (Franks and Fletcher 1983). To achieve this, plants tend to spiral new leaves at a certain angle, and statary army ant colonies seem to follow a somewhat similar strategy to increase hunting efficiency. Studying *Eciton burchellii* on Barro Colorado Island, for example, Franks and Fletcher (1983) found that the raids on subsequent days during the statary phase are rotated around the bivouac at a mean angle of 123°, rather than 90°, which one would expect if raids on subsequent days were random in their direction.

Similarly, during the nomadic phase colonies could avoid plundering the same area repeatedly by raiding in the opposite direction of the previous day's raid. Because colonies emigrate in the direction in which they raid, the consequence would be that consecutive emigrations should approximate a straight line. Franks and Fletcher (1983) indeed presented evidence for such a pattern in their study of *Eciton burchellii*. Throughout a nomadic phase, colonies followed roughly the same compass bearing in their emigrations, and this pattern was only interrupted when colonies failed to emigrate. Accordingly, with an average of 529 meters, the distance between subsequent statary bivouacs was significantly greater than what one would expect if colonies simply moved through the forest in a random walk (Franks and Fletcher 1983). An extensive study of *Eciton burchellii* in the Peruvian Amazon later obtained results seemingly supporting the general conclusions of Franks and Fletcher (1983) with respect to the behavior of both statary and nomadic colonies (Willson et al. 2011).

The ability to avoid previously raided areas is of course contingent on some sort of "memory" of previous raid directions. Franks and Fletcher (1983) suggested that pheromone trails might be important in this respect, and leftover trail pheromones from the previous day could simply demarcate a no-go area. However, as pointed out by Califano and Chaves-Campos (2011), this assumption is problematic. *Eciton burchellii* colonies establish their new bivouacs along the day's raiding trail, but not necessarily at the most distal point of the trail. Trail pheromones, on the other hand, only indicate the location of the trail, but not its direction. As a new swarm raid leaves the bivouac at dawn, the ants should thus perceive trail pheromone in two directions and be oblivious to the trail's directionality. Indeed, in their study at La Selva Biological Station, Califano and Chaves-Campos (2011) showed that *Eciton burchellii* colonies do not avoid trail pheromone as the swarm raid develops. However, in that population, colonies also do not seem to maintain the same compass bearing throughout the nomadic phase (Califano and Chaves-Campos 2011); that is, they differ in this respect from colonies from Barro Colorado Island and Peru (Franks and Fletcher 1983; Willson et al. 2011). Similarly, Mirenda and Topoff (1980) failed to find correlations between the directions of subsequent emigrations in *Neivamyrmex nigrescens*.

Why would such a seemingly important aspect of *Eciton burchellii* biology vary so dramatically between populations? And what are the differences in sensory biology that underlie this striking difference in

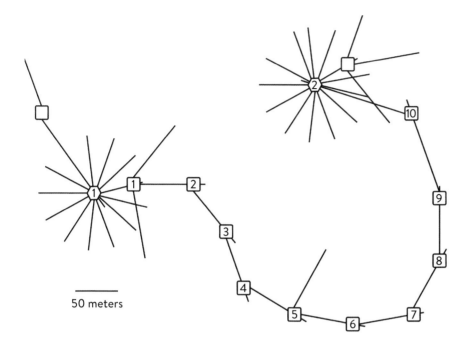

FIGURE 4.27 Schematic of the raiding and emigration activity of *Eciton burchellii* during the colony cycle. Statary and nomadic bivouacs are depicted as hexagons and squares, respectively. The colony leaves statary bivouac 1 at the beginning of the nomadic phase and enters statary bivouac 2 at the end of the nomadic phase. Raids and emigrations are depicted as lines. The average statary phase lasts for twenty-one days, but the colony conducts raids on only some of those days. The average nomadic phase lasts for fourteen days. The colony raids every day during the nomadic phase, but it may skip a few emigrations. In particular, colonies tend to stay longer at the first nomadic bivouac. Emigrations occur along the main raiding column of a given day but usually do not cover the full distance of the raid. The first emigration of the nomadic phase is often comparatively short, while especially the last emigrations tend to be longer. The schematic is loosely based on data from Teles da Silva (1977a, 1977b, 1982).

behavior? These questions can only be answered with additional research, including a careful reanalysis of the existing data. Finally, colonies should not only avoid repeatedly raiding the same area, but they should also avoid areas that have recently been raided by other colonies. Franks and Fletcher (1983) therefore suggested that the trail pheromone of army ants might be long-lasting, and colonies could thus detect areas that have been visited not long ago. Abandoned trails of *Eciton burchellii* and *Eciton hamatum* are indeed still detectable by the ants after a few days (Torgerson and Akre 1970). However, the available empirical evidence in fact suggests that *Eciton burchellii* colonies do not avoid areas previously raided by other colonies (Willson et al. 2011). Some of the behavioral aspects of the phasic army ant colony cycle discussed here are summarized in Figure 4.27.

Interestingly, as was first noted by Müller (1886), the presence of different brood developmental stages throughout the colony cycle correlates strikingly with the cyclic changes in colony behavior. In the presence of pupae, the bivouacs tend to be more sheltered and more persistent. This is of course true for the statary bivouacs of both *Eciton burchellii* and *Eciton hamatum,* but also applies to a lesser extent to the first bivouac of the *Eciton burchellii* nomadic phase. When the colony contains larvae in mid-development, on the other hand, bivouacs are exposed and usually only last for a single night. In fact, as we shall see in the following section, it turns out that cues derived from the developing brood are key regulators of the army ant colony cycle.

The Proximate Causes of Phasic Behavior

As the ethologist Niko Tinbergen so famously pointed out, an integrative understanding of animal behavior requires both proximate and ultimate explanations. Proximate explanations provide a mechanistic understanding of behavior by elucidating, for example, the environmental or physiological factors that immediately regulate the behavior. Ultimate explanations, on the other hand, shed light on the adaptive value of a behavior by emphasizing the selective pressures that have given rise to the behavior over evolutionary time. In this section, we will evaluate the proximate factors that regulate army ant colony cycles, and in the final section of this chapter we will discuss potential ultimate factors that have led to the evolution of regular colony cycles in some but not all army ants.

Most organisms are subject to some form of biological rhythm. Many of these rhythms are controlled by an external *Zeitgeber,* a pacemaker that sets the period of the rhythm. Rhythms with an external *Zeitgeber,* called exogenous rhythms, are animal migrations that occur in response to changes in the seasons, or the lunar cycles of many marine organisms that synchronize their reproduction with the phases of the moon. Biological rhythms that are mainly controlled by factors internal to the animal are called endogenous rhythms, such as female menstrual cycles or the circadian clock. The army ant colony cycle is peculiar because, as it turns out, it is endogenous at the level of the colony superorganism, while it is exogenous at the level of the individual ant.

The cycles of different colonies in a given population of *Eciton burchellii* or any other phasic army ant are not synchronized, as would be expected if the phase of the moon or some other external factor affecting all colonies equally were acting as a pacemaker (Schneirla 1971). Nor are the colony cycles subject to an endogenous reproductive cycle of the queen (Schneirla 1971).

Wilhelm Müller was the first to realize that the changes in army ant colony behavior seemed to be functionally related to brood development, a conjecture that was based on a single observed transition from nomadic to statary behavior (Müller 1886). Theodore Schneirla, who later conducted long-term field studies on a number of phasic army ants, including *Eciton burchellii, Eciton hamatum, Neivamyrmex nigrescens,* and the Asian *Aenictus laeviceps,* developed what he called the "brood-stimulative theory" (Schneirla 1971, 151): "Briefly, as each statary phase ends, the colony is aroused by high-level stimulation from an emerging callow brood and begins a nomadic phase. As the callow-arousal effect wanes, the nomadic phase is maintained by the stimulative effect of a developing larval brood. When these larvae become mature, the colony lapses into reduced function, and a statary phase then runs its course under low-level stimulation from this brood in its prepupal and pupal stages."

Schneirla's research on army ants undoubtedly uncovered an impressive amount of biological detail, but his brood-stimulative theory rested almost entirely on observed correlations between the reproductive physiology of the queen, brood development, and colony behavior, rather than on experiments that could have helped establish causality and disentangle how exactly different factors give rise to the army ant colony cycle (Sudd 1972; Hölldobler and Wilson 1990). In other words, the evidence brought forward by Schneirla was largely circumstantial. And despite the fact that Schneirla's observations suggested a series of obvious experiments involving manipulations of the presence of different brood developmental stages in the colony, Schneirla made no apparent effort to conduct such studies (Sudd 1972). Under more rigorous modern scientific standards, one might thus regard Schneirla's theory as little more than a plausible hypothesis that remains to be tested.

In Schneirla's defense, experimenting with the large and aggressive colonies of *Eciton* army ants in the field is essentially impossible. Following up on the work of his former advisor, Howard Topoff, a student of Schneirla, subsequently undertook some brave attempts to work with the somewhat smaller colonies of *Neivamyrmex nigrescens,* experimentally manipulating them both in the field and in the laboratory (Topoff and Mirenda 1980b; Topoff et al. 1980b, 1981; Topoff 1984). The challenges of working with a real army ant, however, persisted. Sample sizes in these experiments were often small, important environmental parameters difficult to control, and results therefore not easily interpretable.

But the study of doryline colony cycles has recently undergone a renaissance with the emergence of a peculiar laboratory model system, the clonal raider ant *Ooceraea biroi,* which we met briefly in Chapter 2 (see Table 2.2 and Figures 2.24 and 2.25). The first report on the biology of this species was published in 1995 by the Japanese biologists Kazuki Tsuji and Katsusuke Yamauchi (1995). The clonal raider ant belongs to the subfamily Dorylinae and is a subterranean predator of other ants, just like most army ants. Over the course of a few years, Tsuji and Yamauchi were able to collect several colonies of this enigmatic species on the Ryukyu Islands and maintain them in the laboratory inside simple nest boxes. The colonies they had encountered in Japanese soil ranged from only 150 to 600 workers, three to four orders of magnitude smaller than the colonies of *Eciton* or *Dorylus* army ants.

In the laboratory, Tsuji and Yamauchi made a series of intriguing observations. First, colonies contained no queens, only female workers that seemed to produce new workers asexually, a process known as thelytokous parthenogenesis. As it later turned out, the workers literally clone themselves, so individuals in a clonal raider ant colony are almost genetically identical (Kronauer et al. 2012, 2013). But Tsuji and Yamauchi also noticed that the developmental stages in clonal raider ant colonies seemed strikingly synchronized, and that workers became less active and formed a tight cluster once the larvae had pupated, suggesting that colonies might in fact undergo phasic changes in reproduction and behavior.

The first detailed description of the clonal raider ant colony cycle was published a few years later by two researchers from the University of Paris, Fabien Ravary and Pierre Jaisson (2002), showing that indeed even small laboratory colonies undergo regular cycles much like the enormous colonies of *Eciton* and other phasic army ants (Figure 4.28).

FIGURE 4.28 Just like the colonies of many army ants, the colonies of several non–army ant dorylines, such as the clonal raider ant *Ooceraea biroi* shown here, undergo phasic colony cycles (see Table 2.2). However, unlike army ants, clonal raider ants can easily be kept in the laboratory, and their colonies can be experimentally manipulated. On the left is a laboratory colony in the brood care phase. In this phase of the cycle, the colony only contains larvae, which are fairly synchronized in their development. The same colony in the reproductive phase containing eggs and pupae is on the right. This species is unusual in that it does not have queens, and colonies consist only of workers, all of which can reproduce asexually. The colony cluster in the brood care phase contains slightly fewer workers because some have left the colony to forage. In clonal raider ants, foraging ceases entirely during the reproductive phase. Laboratory colony collected on St. Croix, U.S. Virgin Islands, and maintained at the Rockefeller University.

Finally, researchers were poised to experimentally establish the proximate causality behind army ant colony cycles. In a classic follow-up paper, Ravary and Jaisson performed the elegant experiments that had eluded earlier researchers, adding or removing different brood developmental stages from colonies at different stages of the cycle, while carefully monitoring how their manipulations would affect the reproductive physiology and behavior of the ants (Ravary et al. 2006).

From these and subsequent experiments with the clonal raider ant has emerged a model in which the doryline colony cycle can be understood as an oscillator arising from a simple negative feedback loop (Figure 4.29). Oviposition, the laying of eggs, occurs in the absence of larvae, and once the larvae have hatched from the eggs, they soon exert two main effects on the adult ants. First, they induce brood care and foraging behavior. In fact, larvae regulate adult activity in a dose-dependent manner, which might account for even the more subtle behavioral changes of an *Eciton* colony as larvae grow during the nomadic phase (Ulrich et al. 2016). Second, they suppress further egg-laying; egg-laying resumes only once the larvae have pupated (Ravary and Jaisson 2002; Ravary et al. 2006; Teseo et al. 2013; Ulrich et al. 2016). Interestingly, workers respond to the presence of larvae by reduced insulin signaling. Insulin signaling is a conserved metabolic pathway that integrates information about the nutritional state of an organism to modulate food intake and reproduction, among other things. In clonal raider ants, it has been placed under social control and regulates the phasic colony cycle (Chandra et al. 2018; Libbrecht et al. 2018). Extrapolating from the clonal raider ant, we may hypothesize that the dramatic physiological changes in *Eciton* queens are induced by larval cues in much the same way as larvae regulate the transitions between brood care and reproduction in the clonal raider ant: by acting on insulin signaling (see Figure 4.29).

Other factors can of course modulate the cycle phenotype and the precise timing of events in a species-specific manner, but these appear to be secondary in nature. For example, although the arousal of workers by their youngest nestmates played an integral part in Schneirla's brood-stimulative hypothesis, Ravary and colleagues (2006) showed that the initiation of the nomadic phase does not require the eclosion of

FIGURE 4.29 The phasic colony cycle of the clonal raider ant, a non–army ant doryline, arises from a negative feedback loop in which the larvae repress egg-laying in the adults, probably via pheromone signaling (Ravary and Jaisson 2002; Ravary et al. 2006; Teseo et al. 2013; Chandra et al. 2018; Libbrecht et al. 2018). It is thus conceivable that the same negative feedback loop regulates the reproductive cycles of phasic army ant queens. This hypothesis, which awaits experimental verification, is depicted in this flowchart.

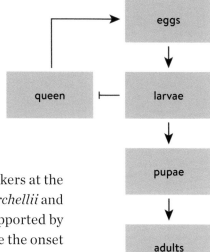

a cohort of callow workers. That the emergence of callow workers at the transition from the statary to the nomadic phase in *Eciton burchellii* and some other phasic army ants is rather coincidental is also supported by the fact that the cohort of callows emerges several days before the onset of the nomadic phase in species like *Aenictus laeviceps* (Schneirla and Reyes 1969; Schneirla 1971). Thus, the emergence of callow workers is neither necessary nor sufficient to induce and sustain a nomadic phase.

How exactly the developing larvae exert these behavioral and physiological effects on their queen mother and their worker sisters remains to be determined. Schneirla (1971) stressed the apparent food begging behavior of the larvae, and indeed army ant larvae are unusually mobile and active compared with the larvae of most other ants. The intensity of their wiggling around might well signal their request for attention and food to the workers. At the same time, the sensory world of army ants is dominated by olfaction and gustation, and it thus seems plausible that larvae would also communicate their needs via chemical cues, or pheromones.

The presence of brood pheromones in ants is still a somewhat contentious issue that has not been fully resolved (Morel and Vander Meer 1988). However, circumstantial evidence suggests that brood pheromones might indeed play a role in regulating phasic army ant colony cycles. In particular, during her morphological and histological studies of larval development in *Eciton burchellii,* Eleanor Lappano, also a student of Schneirla, noticed that the development and activity of the larval labial gland correlates strikingly with the activity pattern of the colony. As the name suggests, the labial gland opens on the labium, the floor of the insect mouth, and it can secrete chemicals that could then, in principle, be sensed or taken up by workers while feeding or transporting the larvae. The gland is inactive in very young larvae present toward the end of the statary phase, but it becomes active in the largest larvae around the transition from statary to nomadic behavior. As more and more larvae begin to secrete labial gland compounds, the raiding and nomadic activity of the colony increases. At the very end of the nomadic phase, the function of the larval labial gland changes: instead of secreting the putative brood pheromone, it now produces and accumulates the precursor of the silk that the larvae will spin into their pupal cocoons. As the larvae gradually undergo this developmental transition, the colony becomes less active and enters the next statary phase (Lappano 1958; Wang and Happ 1974).

In honeybees, the larval labial gland is indeed the source of a brood pheromone, composed of a complex mix of different volatile and nonvolatile chemicals (Le Conte et al. 1990; Slessor et al. 2005; Maisonnasse et al. 2010). Exposure to the pheromone prevents worker bees from laying eggs and increases their foraging efforts (Slessor et al. 2005; Maisonnasse et al. 2010). Thus, the physiological and behavioral effects of brood pheromone on honeybee workers are somewhat similar to those postulated for army ants. However, whether army ant larvae indeed produce a brood pheromone that regulates cyclic colony behavior, and what the glandular origin and chemical basis of this putative pheromone could be, will have to await further research in the future.

The Ultimate Causes of Phasic Behavior

Studying the adaptive value of a behavioral or morphological trait—in other words, inferring the selective pressures that gave rise to the trait— is challenging because evolutionary trajectories usually cannot be observed directly. Therefore, inferences have to be grounded in data that can be collected by present-day investigators. But evolutionary biologists are savvy historians who can employ a series of complementary approaches to shed light on the problem of ultimate causality.

The first approach is the comparative method, which asks whether the trait of interest has evolved repeatedly under specific conditions, thereby pointing investigators toward the causality of evolutionary sequences. As we saw earlier and in Chapter 2, phasic colony cycles occur widely throughout the ant subfamily Dorylinae, but no formal analyses have been conducted to decide whether the trait has evolved only once or multiple times independently, and whether it has been secondarily lost in some instances (see Tables 2.2 and 2.3). Phasic colony cycles might also have evolved independently at least once in the ponerine genus *Simopelta* (Gotwald and Brown 1966; Kronauer et al. 2011a). In the leptanilline genus *Leptanilla* (Masuko 1990) and the amblyoponine genus *Onychomyrmex* (Miyata et al. 2003), synchronized broods have been reported, but this seems to reflect a seasonal phenomenon rather than a phasic colony cycle as in many dorylines (see Table 2.1). Either way, even though not all army ants are phasic, phasic colony cycles have only evolved in species with army ant–like biology. This suggests that a phasic lifestyle is indeed adaptive in the specific context of social predation, at least under some circumstances (Garnier and Kronauer 2017).

A defining feature of army ants is that they hunt in large groups (Chapter 3). Nomadism (this chapter) and reproduction by colony fission (Chapter 5), the other two pillars of the army ant adaptive syndrome, seem to have facilitated the evolution of obligate mass raiding (Chapter 2). My colleague Simon Garnier and I therefore employed a second approach, constructing an explicit theoretical model to explore the possible link between phasic colony cycles and army ant foraging (Garnier and Kronauer 2017; see also Teseo and Delloro 2017). We reasoned that most army ants and their relatives would face a "high cost of entry" scenario when foraging. Under this scenario, a small foraging party would be inefficient because it would be unable to overwhelm prey items such as large arthropods or the well-fortified colonies of social insects. However, at a substantial colony investment into foraging, the relative returns should be high and therefore allow the ants to feed a large number of larvae. In contrast, most other ants should face a "resource exhaustion" scenario. Under this scenario, foraging becomes relatively more costly as the number of ant larvae that have to be fed increases, because foragers have to venture farther and farther from the nest to find food. Our model showed that under resource exhaustion colonies work most efficiently by keeping the food demand of the larvae at a constant intermediate level because this avoids the high relative foraging costs associated with large numbers of larvae. Under high cost of entry, on the other hand, colonies perform best by alternating between states where they either forage excessively or not at all, thereby avoiding the disproportionally small returns associated with foraging at low levels. This is exactly what phasic army ant colonies achieve by developmentally synchronizing large brood cohorts.

All army ants are nomadic by definition, but not all army ants show phasic behavior (see Table 2.3). For example, no regular pattern appears to underlie the frequent colony relocations of species in the genus *Dorylus,* and brood of all developmental stages can be found in a given colony when excavated in its entirety (Raignier and van Boven 1955; Schneirla 1957a, 1957b, 1971; Raignier et al. 1974; Leroux 1982; Berghoff et al. 2002b; Schöning et al. 2005b) (Figure 4.30). Unfortunately, unlike *Eciton burchellii* and *Eciton hamatum, Dorylus* colonies, like most army ants in fact, construct their bivouacs entirely below ground and are thus less amenable to observation. The same applies to their emigration columns, which even in the most surface-adapted species, the driver ants, are only partially visible above ground and usually covered by earthen tunnels (Figure 4.31). Data on those species are thus more difficult to collect and therefore comparatively sparse.

As a graduate student at the Free University of Berlin, Caspar Schöning, who would later become my collaborator on the study of African driver ants, undertook a heroic effort to monitor colony emigrations in *Dorylus molestus* at Mount Kenya. He found that colonies can stay at a given nest site anywhere between three and 111 days before moving on, confirming previous reports that the emigrations of *Dorylus* army ants do not follow any apparent temporal pattern (Schöning et al. 2005b; see also Raignier and van Boven 1955; Schneirla 1971; Leroux 1982). The average distance of a *Dorylus molestus* emigration was 93 meters in that study, and others have reported mean distances of up to 233 meters for driver ant emigrations. Because of the massive size of driver ant colonies, these emigrations can take up to five days and nights to complete (Gotwald 1984–1985).

FIGURE 4.30 The colonies of *Dorylus* army ants do not undergo phasic cycles, and brood development is not synchronized to the extent observed in phasic species like *Eciton burchellii* and *Eciton hamatum*. This snapshot of a *Dorylus molestus* emigration at Mount Kenya shows workers carrying a wide range of developmental stages, including eggs, larvae, and pupae. Unlike in *Eciton* army ants, the pupae of *Dorylus* army ants are not enclosed in cocoons.

FIGURE 4.31 Driver ants, here the species *Dorylus wilverthi*, relocate their colonies largely below ground, and wherever their emigrations are forced to surface the ants construct earthen galleries. Shown here is a short stretch where the tunnel opens into a half-pipe, and the emigration is heavily guarded by the largest workers. These workers functionally correspond to the distinct soldiers of *Eciton* army ants and are overrepresented wherever the colony is most vulnerable to attack (Braendle et al. 2003). In the background, a worker is carrying a white driver ant pupa. Kakamega Forest, Kenya.

What triggers the irregular emigrations of nonphasic army ants is not entirely clear. In the ponerine *Leptogenys distinguenda,* food supplementation decreases both raiding and emigration activity, and it has been suggested that colonies emigrate if they find a suitable nesting site during an extensive raid that leads into a new, resource-rich patch (Witte and Maschwitz 2000). In *Dorylus molestus,* similar feeding experiments resulted in shorter emigrations, but had no effect on how frequently colonies emigrated (Schöning et al. 2011). Instead, at least in some cases, driver ant emigrations might be triggered by encounters with competing colonies, predator attack, or unhygienic nest conditions (Schöning et al. 2011).

This leaves us with the question of whether nonphasic army ants such as *Dorylus* do not face a high cost of entry foraging scenario, as would be predicted by our model (Garnier and Kronauer 2017). We have argued that this might in fact be the case. As discussed in Chapter 3, the prey spectrum of *Dorylus* army ants is unusually broad. While most army ants prey exclusively or predominantly on other social insects, most *Dorylus* species are far less selective when it comes to live prey. Many species will even feed on animal carcasses, nuts, fruits, grains, and vegetable oil, food items that other army ants will not touch (Gotwald 1995; Berghoff et al. 2002a). This suggests that food is more readily available for *Dorylus* army ants, and that even modest foraging parties might be efficient, especially when feeding on plant material or preying on small solitary arthropods.

The third approach to test the plausibility of adaptive scenarios is direct experimentation. To further evaluate some of the conjectures put forward in this section, one could disrupt the cycle of phasic colonies over an extended period of time and record whether and why exactly those colonies perform poorly in comparison to their unperturbed peers. One could also attempt to measure the actual relationship between larval food demand and foraging costs by experimentally manipulating the number of larvae inside colonies of phasic and nonphasic dorylines. Although this approach might be feasible in species like the clonal raider ant, the prospect of manipulating larval cohorts and measuring foraging effort and efficiency in *Eciton* or *Dorylus* army ants seems daunting and painful to say the least. Clearly, our understanding of the proximate and ultimate factors giving rise to the peculiar colony cycles of phasic army ants has progressed substantially over the past two decades, and several promising yet unexplored avenues for further research remain.

In this chapter, we have discussed how army ant colonies, like giant nomadic tribes, roam the tropical rainforests of the world. This lifestyle is probably ultimately dictated by their voracious appetite for live prey and the associated necessity to frequently relocate to new hunting grounds. But relying entirely on predation in large groups not only meant that army ant colonies had to forego a permanent abode, they also could no longer afford to go through the incipient stages of colony foundation that are typical for most other ants. This dilemma required yet another major innovation in ant life history: reproduction by colony fission. This third pillar of the army ant adaptive syndrome will be the subject of the following chapter.

CHAPTER FIVE

Colony Fission

The birth of a new army ant colony is a spectacular and rare event. Monica Swartz, who spent hundreds of days tracking *Eciton burchellii* colonies in Costa Rica's Corcovado National Park as a doctoral student, observed only a single colony reproduce during that time (Swartz 1997). Theodore Schneirla, who conducted extensive fieldwork on Barro Colorado Island in Panama, encountered twenty-three instances in which an *Eciton* colony was raising new queens and males, out of 360 broods surveyed across all *Eciton* species present on the island (Schneirla 1971). When an *Eciton burchellii* colony approaches fission—the splitting of one large colony into two—it raises a brood of up to six young queens and about 2,500 to 4,000 males, but no workers. The smaller *Eciton hamatum* colonies raise about 1,500 males and the same number of queens as *Eciton burchellii* (Schneirla 1948, 1956, 1971; Schneirla and Brown 1950; Franks 1985; Franks and Hölldobler 1987; Kronauer et al. 2007b). So if you are lucky enough to encounter an *Eciton* colony in the first place, the probability that it is in the process of raising such a sexual brood is somewhere on the order of 6 percent, according to Schneirla's data. And even if you win the army ant lottery, you will usually have to follow the colony for another few weeks, until the beginning of the next nomadic phase, which is when the ceremonious act of colony fission will take place.

On March 13, 2017, Philipp Hönle, then a student with my colleague Christoph von Beeren, came across such a colony of *Eciton burchellii* at La Selva Biological Station. He encountered the colony at around 3:00 p.m., when it was already emigrating. This timing in itself is unusual for an emigration of this species, but the size of the colony must have been truly extraordinary: at 2:00 a.m. the next morning, when Phil decided it was finally time to go to bed, the procession was still underway, with no end in sight. On March 17, Phil sent me a few photographs to confirm what he had already suspected: the colony was raising far fewer larvae than usual, but the larvae the workers regularly schlepped to new bivouacs were gigantic. Royalty in the making. Once the larvae had pupated and the colony had settled for the statary phase between the roots of a large rain-forest tree, we were able to predict the onset of the next nomadic phase and, therefore, the upcoming colony division. It was now time to book my flight to Costa Rica. I had followed a colony with a sexual brood in Venezuela previously, but that colony did not divide in the end (Kronauer et al. 2007b). Therefore, what was about to unfold at La Selva would be a spectacle that I certainly didn't want to miss.

In contrast to the sheltered statary bivouacs of *Eciton burchellii* colonies raising worker broods, discussed in Chapter 4, those containing sexual broods tend to be more exposed and often visible, although the evidence for this being a general feature is of course rather anecdotal (Kronauer et al. 2007b). Nevertheless, the statary bivouac of this particular colony fit this pattern, and at 4:00 p.m. on April 11, I observed the first discarded male pupal cases on the bivouac surface (Figure 5.1). The males had begun emerging from their cocoons, and colony fission was now imminent. By 7:00 p.m., a dense emigration trail had formed to a new nascent bivouac, located between the buttress roots of a tree giant about 10 meters away. At 8:00 p.m., the old and still partially physogastric queen commenced her journey toward this new roadhouse. Males, however, were still nowhere to be seen.

After an otherwise uneventful night, I returned to the bivouacs at 8:30 a.m. the next morning to find that they were still connected by strong, bidirectional traffic. But in addition to this umbilical cord, both bivouacs had developed their independent raiding columns, one emanating from the new site, and two from the large statary bivouac. What's more, the surface of the statary bivouac was now covered in large empty pupal cases, even though the males themselves remained elusively hidden inside the fortress of workers. However, I finally caught a glimpse of the

FIGURE 5.1 Empty pupal cases from a sexual brood are visible on the surface of an *Eciton burchellii* bivouac at the end of the statary phase. Unlike *Eciton burchellii* statary bivouacs with worker broods, those containing male and queen pupae are often exposed.

first young queen, traveling outbound on one of the independent trails leaving the statary bivouac. Colony fission was now in full progress.

Several hours later, at 4:00 p.m., I discovered the first male, still inside its intact cocoon, being carried along one of the emigration trails by his sister workers. The workers' excitement during emigrations with sexual broods or males is so elevated that they form exceptionally broad columns that are literally paved with the bodies of live ants, a phenomenon that Schneirla described as "ant roadways" (Schneirla 1971) (Figure 5.2). With the density of traffic also increases the ants' road rage, making close-up observations particularly daunting.

Between 5:40 and 5:50 p.m. I witnessed yet another remarkable behavior. Within a window of just ten minutes, five nonresident *Eciton burchellii* males landed dead on target and with a loud buzz, right on the roots and tree trunk surrounding the statary bivouac. As was immediately evident from their darkened cuticle, these males were older than the ones inside the bivouac and had found their way through the rainforest on the wing, possibly from miles away, to mate with the young queens. How exactly they accomplish this feat is entirely unknown, but the fact that males do not usually target colonies under other circumstances suggests the involvement of some potent pheromone that is somehow associated with the presence of virgin queens. By now the various emigrations were in full force, with a few remaining male cocoons being carried by workers, and recently eclosed males walking along all three columns, often with one or two large workers or submajors attached to them (Figures 5.3 and 5.4). One trail led to the colony fragment with the old mother queen, which was again on the move, while the other two traversed the forest toward more distant bivouacs, each headed by a single young sister queen, as we were able to confirm later. In contrast, in all previously reported cases, fissioning *Eciton* colonies split into exactly two daughter colonies (Schneirla 1971; Franks and Hölldobler 1987). The few supernumerary queens are sealed off by clusters of workers in this process and ultimately are left behind to die. It is of course impossible to know, but maybe it was the truly gargantuan size of this particular colony that let the dynamics underlying colony fission play out slightly differently, ultimately leading to three rather than two fission products.

FIGURE 5.2 When *Eciton burchellii* colonies contain a sexual brood, their emigration columns become even wider and busier, turning into "ant roadways" that are literally paved with the bodies of live ants.

FIGURE 5.3 The nomadic phase of *Eciton burchellii* colonies begins when larvae hatch from the eggs and callows eclose from the pupae. However, during the first emigrations, colonies often still contain a few eggs and pupae that are carried by the workers. In this picture, several workers are teaming up to haul a large male pupa in the first nomadic emigration while the colony is undergoing fission. In the foreground, a worker is carrying a clump of eggs and early instar larvae, which are part of the next worker brood. Toward the middle and end of the nomadic phase, later larval instars will be carried one by one.

FIGURE 5.4 During the first few nomadic days after colony fission, *Eciton burchellii* males are still walking in the nightly emigrations, often being "assisted" by submajors and large regular workers. Once the males have matured, they will climb the low vegetation and take off on the wing, on their quest to find a young queen to mate with.

Raising a Sexual Brood

While the queens of most ants found colonies independently, dependent colony founding has evolved repeatedly throughout the ants (Cronin et al. 2013). In species with dependent colony founding, mature colonies produce propagule colonies with one or several queens that are accompanied by workers and disperse on foot. In many species with dependent colony founding, colonies contain several reproductively active queens. In these polygynous species (from Greek, meaning "many females"), offspring colonies usually remain close to their mother colony. Colony reproduction in these cases is often referred to as "budding." Other species with dependent colony founding, however, are monogynous, with a single egg-laying queen per colony, and offspring colonies disperse over larger distances. These species are said to reproduce via colony "fission." Colony fission is fairly rare among the ants, but it is practiced by all army ants. It is important to keep in mind, however, that, as so often in biology, these are not two discrete categories, and a whole range of dependent colony founding strategies exists (Cronin et al. 2013). For example, as we will see, at least one army ant species is polygynous, yet offspring colonies do not remain in the vicinity of their mother colony.

One striking feature of species with dependent colony founding is that numerical sex ratios—in other words, the numerical ratio of young males to young queens raised by colonies—is male biased, and often, like in army ants, dramatically so, as illustrated by the previously provided numbers for *Eciton burchellii* and *Eciton hamatum*. Even though in the African driver ant *Dorylus wilverthi* the number of males in sexual broods—ranging from several dozens to approximately 3,000—can be more variable and the number of young queens can be more than ten, the same principle applies, especially considering that the workers ultimately seem to kill the supernumerary queens (Raignier 1972; Raignier et al. 1974).

From an evolutionary perspective, this might seem surprising at first. According to Fisher's sex ratio principle, named after Ronald Fisher, one of the founding fathers of modern evolutionary biology, the ratio between males and females produced in a sexually reproducing species should be approximately one to one. The reason is that, as one sex becomes more common, the relative value of the other sex increases because individuals of that sex will produce more offspring on average, given that each offspring has exactly one parent from each sex. Any genetic variant that biases the sex of one's offspring toward the rarer sex will therefore be advantageous and increase in frequency in the population, until the evolutionarily stable Fisherian sex ratio of one to one is restored. However, this simple calculation is not valid in species where individuals of one sex are more costly to produce. This is known as the investment sex ratio, as opposed to the numerical sex ratio. If, for example, females are considerably larger than males and therefore require five times the amount of resources from their parents, then the stable numerical sex ratio should be closer to five males for each female in order to give identical fitness returns for each sex.

Army ant queens obviously are not several thousand times larger than army ant males—in fact, in army ants the two sexes are relatively similar in size (Franks and Hölldobler 1987). However, while males simply disperse from the natal colony, young army ant queens leave with an enormous dowry: they take with them a large proportion of the worker force and the brood. The precise value of this dowry is of course difficult to calculate, but it has to be penciled in when estimating the investment a colony makes to produce a daughter queen (Hamilton 1975; Macevicz 1979; Pamilo 1991; Bourke and Franks 1995; Crozier and Pamilo 1996). Clearly, that investment is immense. Another way to think about this problem is that, under colony fission, there is no point in producing more than a few young queens. Given the constraint of having a single queen per colony, there will be intense local competition between the young daughter queens. Ultimately, only one or, in cases where the mother is superseded, two young queens will end up heading new colonies. Males, on the other hand, compete with other males to mate with young queens at the level of the entire population. It therefore makes sense to produce as many males as possible (Bulmer 1983; Pamilo 1991; Bourke and Franks 1995; Crozier and Pamilo 1996). In other words, while only one or two daughters can become heir to your own throne, every additional son is another ticket in the lottery to inherit a slice of a foreign army.

But which parameters determine whether a colony raises a sexual brood, and who in the colony makes the ultimate decision? What exactly triggers the switch from producing worker broods to producing a sexual brood is unknown. On Barro Colorado Island, the chance of finding sexual broods of *Eciton burchellii* and *Eciton hamatum* is higher in the dry season, and Schneirla (1971) proposed that the dry weather triggered the production of sexual broods via a direct effect on the queen's reproductive physiology. The evidence for such a direct physiological effect is weak, however. Furthermore, not all army ants adhere to the same pattern; in some cases, even closely related species do not overlap in their seasonal patterns at all. For example, while males of *Labidus coecus* can be found at light traps from June to September in the Atlantic Forests of Brazil, *Labidus praedator* males show up between October and May (do Nascimento et al. 2004, 2011; see also Baldridge et al. 1980). And even though African *Dorylus* species show some seasonality in flight activity, males of many species can be encountered throughout the year (Raignier 1972; Raignier et al. 1974; Leston 1979). These patterns suggest that dry weather conditions are not a universal trigger of army ant colony reproduction.

What is clear, however, is that colonies have to grow to a certain size before it makes sense for them to divide again (note, however, that in exceptional cases *Dorylus* driver ant colonies can undergo consecutive divisions; Leroux 1979a). According to one model, colonies should undergo fission when they reach the size at which the combined growth rate of their daughter colonies begins to exceed their own growth rate (Franks 1985). At sites with pronounced seasonality, prey abundance and colony growth rate may vary across the year, which in turn might make it more likely for colonies to reach large sizes and undergo fission during certain months.

Who determines that the conditions are right for a colony to divide? Is it the queen who decides to lay an extraordinary batch of eggs, or is it the workers who decide to raise an extraordinary cohort of larvae? While hymenopteran females arise from fertilized eggs and are diploid, containing two sets of chromosomes, one from each parent, males develop from unfertilized eggs and are haploid, having only a single complement of chromosomes inherited from their mother. Therefore, the queen will have to produce at least a few thousand unfertilized haploid eggs that can give rise to the males of an army ant sexual brood. But that does not necessarily mean that it is the queen's decision to raise a sexual brood. It is entirely possible that some proportion of eggs are always unfertilized, but that the resulting male larvae either starve if they do not receive the royal treatment, or that the workers cannibalize them at some point in the process of raising a worker brood. This latter scenario would imply that the decision to raise a sexual brood is mostly made by the workers.

Answering the question requires detailed data on the ploidy level of eggs, both preceding the production of worker and sexual broods. Unfortunately, these kinds of data currently do not exist. However, the little information that is available suggests that queens do not simply lay a much smaller batch of mostly haploid eggs that then develop into a handful of queens and a few thousand males. Based on a unique observation from *Eciton hamatum,* it seems that the batch of eggs ultimately giving rise to a sexual brood is in fact initially of similar size to batches giving rise to worker broods (Schneirla and Brown 1952; Schneirla 1971). What happens next is quite remarkable, however: before the first colony emigration of the nomadic phase, the vast majority of eggs are eaten, giving the remaining larvae an enormous early growth boost. Even though it has been suggested that many of the devoured eggs are nonviable, trophic ("alimentary") eggs (Schneirla 1971), it seems unlikely that this scale of egg cannibalism and the precisely controlled outcome in terms of the types and numbers of offspring reared could be controlled solely by what kinds of eggs the queen lays. As a consequence of these dynamics, when the colony finally enters the nomadic phase, the young sexual larvae have already attained a body size that equals the size of worker larvae at maturity (Schneirla 1971).

However, the royal treatment of sexual larvae is far from over with the beginning of the nomadic phase. Although an *Eciton* colony raising a sexual brood deals with only about 2 percent of the number of larvae raised in a worker brood, its raiding activity is even more vigorous, and more prey are retrieved on a daily basis (Schneirla 1971). This is despite the fact that colony emigrations are now less frequent than in regular nomadic phases (Schneirla 1971; according to Raignier 1972, colonies of the African driver ant *Dorylus wilverthi* do not emigrate at all when raising a sexual brood). This suggests that, during any given day of the nomadic phase, a sexual larva receives about fifty times the amount of food that the average worker larva gets to eat. Raised on such a diet, male and queen larvae begin to pupate several days and up to a week earlier than worker larvae, and the nomadic phase of colonies raising a sexual brood is accordingly shorter (Schneirla and Brown 1950; Schneirla 1971; Kronauer et al. 2007b). At pupation, their body volume is almost twenty times greater than that of the largest worker larva at the corresponding developmental stage (Schneirla 1971) (Figures 5.5 and 5.6). Interestingly, the queen larvae are consistently a few days ahead of the males in development and are the first to pupate and, later, to eclose (Schneirla 1971). This overall pattern seems to apply to all studied species of *Eciton, Neivamyrmex,* and *Aenictus* (Schneirla 1948, 1956, 1961, 1971; Schneirla and Brown 1952).

FIGURE 5.5 *Eciton burchellii* workers team up to carry a large sexual larva during an emigration. Henri Pittier National Park, Venezuela.

FIGURE 5.6 An *Eciton burchellii* submajor in an experimental laboratory setup is guarding a male cocoon that has been removed from a statary bivouac. The meconium, which the larva excretes as it enters pupation, is visible through the bottom of the cocoon as a black dot.

This brings us to the important topic of caste development. How is it possible to raise different female larvae that do not systematically differ in their DNA sequence into vastly different types of adults? In *On the Origin of Species* (1859), Charles Darwin actually used African *Dorylus* driver ants (Figure 5.7) to illustrate this extreme case of phenotypic plasticity:

> The reader will perhaps best appreciate the amount of difference in these workers, by my giving not the actual measurements, but a strictly accurate illustration: the difference was the same as if we were to see a set of workmen building a house, of whom many were five feet four inches high, and many sixteen feet high; but we must in addition suppose that the larger workmen had heads four instead of three times as big as those of the smaller men, and jaws nearly five times as big. The jaws,

FIGURE 5.7 Even though the workers of African driver ants form a continuous series, they are highly polymorphic. This panel shows the size range in *Dorylus molestus*, ranging from approximately 2.7 to 12.0 millimeters in total body length (Gotwald and Schaefer 1982). The workers differ not only in morphology, but also in behavior. For example, while the largest workers tend to assume defensive postures along trails, the smallest workers are specialized on collecting the queen's eggs. Images of all workers are to scale. Mount Kenya, Kenya.

moreover, of the working ants of the several sizes differed
wonderfully in shape, and in the form and number of the teeth.
(240–241)

Darwin's example becomes only more impressive when the queen is
considered part of the series (Figures 5.8 and 5.9). Furthermore, the different female castes of army ants not only differ in morphology but also
show striking differences in behavior, physiology, and longevity (Schneirla 1971; Topoff 1971; Gotwald 1995; Schöning et al. 2005a).

In ants, the female phenotype at the adult stage, and therefore caste,
is essentially a function of overall body size, or mass, at pupation
(reviewed in Trible and Kronauer 2017). There are different ways in

FIGURE 5.8 A frontal view of the *Dorylus molestus* queen. A small worker is conveniently posing on her head, illustrating the immense level of polymorphism observed
among the female castes of driver ants. Mount Kenya, Kenya.

which a colony can influence larval size, but one of the most obvious determinants is how much and what the larvae eat. In species with independent colony founding, for example, young incipient colonies that cannot forage efficiently often raise only tiny, or "nanitic," workers that are largely absent from mature colonies of the same species. Probably for the same reason, only large colonies with ample food supplies switch from rearing workers to rearing young queens. As outlined earlier, several factors such as colony size, queen physiology, seasonality, and worker behavior might interact in determining when an army ant colony switches from raising workers to raising queens, but inadvertently the result is that the colony raises a single cohort of individuals that are exceptionally well nourished.

The relationship between the overall size of an individual and any particular phenotype does not have to be isometric, which is the point

that Darwin makes with respect to the size of worker heads and mandibles. These are not only absolutely, but also proportionally bigger in larger workers. This is a well-known phenomenon in allometry, the study of relative scaling of different body parts as a function of overall body size, and it applies to both external and internal anatomical features. How exactly any particular body part scales with overall body size can be quite complex. In *Eciton* army ants, for example, relative brain size not only decreases sharply with body size in general, but soldiers have particularly small peripheral chemosensory centers and mushroom bodies, the higher neural processing centers of the insect brain, even considering their already small brain (O'Donnell et al. 2018).

Another interesting case is mandible growth in species like *Eciton burchellii* and *Eciton hamatum*. The growth of the mandibles in workers scales fairly isometrically with body size until we enter the size range

FIGURE 5.9 Depending on their physiological condition, the bizarre queens of *Dorylus* driver ants measure somewhere between 40 and 63 millimeters in total body length, making them the largest ants on the planet (Raignier et al. 1974). Shown here is a queen of *Dorylus molestus*, along with one of her smallest worker daughters. The worker is carrying a bundle of the queen's eggs. Mount Kenya, Kenya.

that produces soldiers. At this point, mandibles begin to overgrow dramatically. If, however, size is pushed even further to produce queens, the mandibles fail to overgrow to the extent observed in soldiers. The reason probably is that in that size range resources during metamorphosis are increasingly allocated to other organs, such as the ovaries, at the expense of the mandibles (Trible and Kronauer 2017). Similarly, one might postulate that soldiers invest in weapons at the cost of brains. All of this is not to say, however, that army ant females could not be arranged in a continuous series, at least in principle (Trible and Kronauer 2017). While worker–queen intermediates have not been observed and might not occur without experimental intervention, the continuous nature of worker morphology becomes apparent in careful morphometric studies with large sample sizes (see, e.g., Cohic 1948 vs. Hollingsworth 1960; Gillespie and Cole 1950; Schneirla et al. 1968; Topoff 1971; Schöning et al. 2005a; Powell and Franks 2006).

One baffling exception has been reported from the Asian army ant *Dorylus orientalis,* where morphologically distinct worker series co-occur, in at least some colonies (Eguchi et al. 2014). Even though standard DNA barcoding revealed no genetic differences between the two forms (Eguchi et al. 2014), based on what we know about ant caste development it seems unlikely that a single genome could regularly produce such striking morphological differences in workers with broadly overlapping size ranges. This system might therefore turn out to be particularly instructive regarding the molecular underpinnings of caste development and possibly other aspects of sociobiology.

Not only are army ant queens much larger than the workers, they are also remarkable in several other ways. As discussed in Chapter 2, they are permanently wingless, which is why their alitrunk resembles that of ant workers. Because they never fly they do not require the well-developed eyes found in queens of many other ants, and they are similar to workers also in that respect. For these reasons, army ant queens are sometimes referred to as ergatoid, or worker-like. But in other aspects they could not be more different from ant workers. Their gaster is extremely large and capable of massive physogastric expansion, to a level that far exceeds other ants but resembles termite queens instead (see Figures 4.23, 4.24, 5.8, and 5.9). The technical term for this unusual combination of traits is "dichthadiigyny" (Chapter 2). Army ant queens also possess an exuberant repertoire of exocrine glands, as do the queens of army ant-like *Leptanilla* and *Onychomyrmex* (Hölldobler 2016). Although the queens of *Leptogenys* army ants and many non–army ant dorylines are still permanently wingless, they are not dichthadiiform and are not endowed with exocrine glands to the same extent as doryline army ant queens (Hölldobler 2016). As mentioned earlier, the permanently wingless queens of army ants are an evolutionary corollary of colony reproduction by fission and, in the extreme cases of dichthadiiform queens, large colony sizes (e.g., Cronin et al. 2013) (see Chapter 2).

Even though army ant males must undoubtedly also receive an enormous amount of food at the larval stage, the switch from female to male production is of a different nature than that from producing workers to producing queens. Unlike the different female castes, which do not systematically differ genetically, females and males differ dramatically in their genomic content. We know that, as in other Hymenoptera, female army ants are diploid and males are haploid, but how sex is determined at the genetic level in army ants remains unclear. In honeybees, a single, highly polymorphic genetic locus determines the sex of an individual (Beye et al. 2003). If an individual has two different genetic variants, or alleles, at this locus (it is heterozygous) it develops into a female. If the individual is haploid and therefore has only one copy of the gene, it develops into a normal, haploid male. However, if an individual is diploid but happens to have the same allele twice (it is homozygous) it develops into a sterile diploid male.

From an evolutionary perspective, producing diploid males is a waste of colony resources, and honeybee colonies that produce large proportions of diploid males at the expense of diploid workers and queens are not viable (Winston 1991; Palmer and Oldroyd 2000). Even though sex determination loci have not been genetically characterized in ants, the fact that diploid males can be found in many species suggests that the basic mechanism might often be similar (van Wilgenburg et al. 2006). Although diploid males have not been observed in army ants, they are produced in at least one non–army ant doryline, the clonal raider ant

Ooceraea biroi (Kronauer et al. 2006b, 2007b, 2012). The possible fitness costs associated with producing diploid males will become important when we discuss the army ant mating system later in this chapter.

The Process of Colony Fission

The shortened nomadic phase of an *Eciton* colony with a sexual brood is followed by a statary phase of normal length (Schneirla and Brown 1950; Schneirla 1971). At the end of that statary phase, it is finally time for the colony to divide. Colony fission in army ants is a complex process, and given that it is impossible to manipulate it in an experimentally controlled and repeatable manner, the underlying dynamics are not well understood. However, based on detailed observations, mostly by Theodore Schneirla and colleagues, one can at least outline what a representative colony division in *Eciton hamatum* or *Eciton burchellii* looks like (Schneirla 1949, 1956, 1961, 1971; Schneirla and Brown 1950, 1952).

The first signs of division become apparent during the later stages of the nomadic phase during which sexual larvae are being reared (Schneirla 1949, 1971; Schneirla and Brown 1950). In normal circumstances the queen is usually located at the top of the bivouac above the developing worker brood; now she occupies one side of the bivouac while the male and queen larvae are segregated inside the other half. This division becomes especially pronounced during the subsequent statary phase. During and after pupation, the young queens are spaced out in tight clusters of workers at the base of the bivouac. The young queens eclose several days before the males and remain inside clusters of workers at the base of the bivouac or within a few centimeters from the bivouac. Unlike in honeybees, which also reproduce via colony fission (Winston 1991), there is no direct combat between the callow queens. However, there are clear and persistent differences in how many workers are clustered around each queen and how active these clusters are, suggesting that queens might differ in their attractiveness to the workers (Schneirla 1971). Army ant queens are extremely attractive and capable of assembling large worker retinues around them, both inside the bivouac and during emigrations (Watkins and Cole 1966; Rettenmeyer et al. 1978; Franks and Höll-dobler 1987; Hölldobler 2016) (see Figures 4.2 and 4.3). One reasonable hypothesis is that the queens advertise their fertility and thus their suit-ability as a future regent to the workers, and that their exuberant endow-ment with exocrine glands plays an important role in the competition to win the workers' favor (Hagan 1954a, 1954b, 1954c; Whelden 1963; Schneirla 1971; Franks and Hölldobler 1987; Hölldobler 2016).

In *Eciton burchellii,* the queen exodus starts shortly after the males begin to eclose. Usually the old mother queen leaves the nest along a raiding trail on her side of the bivouac first and establishes a new cluster of workers not too distant from the statary bivouac. After that, the young queens begin to push outward along a different raiding trail. The queen that had established the largest worker cluster during the statary phase typically takes the lead and moves fairly unobstructed, while workers seal off the other, apparently less attractive, young queens in tight clus-ters at the bivouac or within a few meters from the bivouac along that same route. Ultimately, these supernumerary queens will be abandoned but normally not killed directly (Schneirla 1971). Even though the old queen usually leaves the bivouac first along one of the raiding trails, she can still end up constrained in a tight cluster of workers in the periphery of the trail. This usually indicates that she will be superseded by one of her daughter queens, and that the fission will give rise to two colonies that are each headed by a callow queen. Queen supersedure seems to happen in cases where the mother queen is old and probably declining in her fertility and attractiveness to the workers. Unlike queen movements during colony emigrations, these dynamics take place during the day while the colony is actively raiding. Meanwhile, the males and the next worker brood remain inside the bivouac (Schneirla 1971).

The actual colony division and emigrations then begin around dusk, just as normal emigrations with callow workers. Once the traffic has become entirely outbound, newly eclosed males begin to walk in both emigration columns, and workers carry the remaining male cocoons and the eggs and young larvae that will give rise to the next cohort of workers in both directions (see Figures 5.3 and 5.4). Similar to worker broods (Chapter 4), in *Eciton hamatum* all males have usually eclosed by the time of colony fission, and no cocoons are being carried (Schneirla 1971). During this first emigration and for the next few nights, submajors often run behind the males, holding onto their folded wings or gasters (see Figure 5.4). The two colonies will remain connected by a thin column of workers for a day or so after fission, before they finally separate completely.

The dynamics underlying colony fission in army ants are complex and can be easily destabilized. For example, experimentally removing queens or even queen larvae at different stages of colony reproduction has repeatedly been associated with a failure of colonies to fission (Schneirla and Brown 1950; Schneirla 1971; Kronauer et al. 2007b). But even when experimentally undisturbed, colonies sometimes fail to divide after having raised a sexual brood; in these cases, the old queen is either retained or superseded by a callow queen. As far as we know, however, some young queens are always produced as part of a sexual brood in *Eciton* (Schneirla 1971). In African driver ants, on the other hand, colonies seem to raise males without new queens quite regularly, even though it is of course difficult to say whether young queens were initially present in the brood but died at some point (Raignier 1972; Raignier et al. 1974).

While the general mode of colony reproduction is always the same, the precise process of colony fission differs in other army ant genera (Schneirla 1971). In Asian *Aenictus,* for example, the males take off at the end of the statary phase before their natal colony undergoes fission (Schneirla 1971). From *Aenictus* also comes the only other report where an army ant colony divided into three offspring colonies, one headed by the old mother queen, and two headed by callow daughter queens (Schneirla 1971). In the African driver ants *Dorylus nigricans* and *Dorylus wilverthi,* the exodus of the old mother queen with a large proportion of the workers and young worker brood leaves behind all of the male brood, which at this point can be larvae, pupae, or young adults, often together with one or more callow queens (Schneirla 1971; Raignier 1972; Raignier et al. 1974).

After colony fission has been completed in *Eciton,* the young males initially remain with their colonies. After three or four nights, however, they begin climbing up on vegetation and vibrating their wings, while excited workers still cling to them. Eventually, more and more males break away from the workers, mostly early in the emigration, and fly off into the tropical sunset, with most of the males leaving about halfway into the nomadic phase (Schneirla 1971). The young queen, on the other hand, stays behind. Based on dissections of *Eciton hamatum* queens early into their first nomadic phase, it appears that callow queens are already inseminated by foreign males either directly after eclosion in the statary bivouac, or shortly thereafter (Schneirla and Brown 1950; Schneirla 1971). By the time they enter their first statary phase as adults, the young queens are able to produce a worker brood that resembles that of an established queen in size (Schneirla 1971).

An important corollary of colony fission and wingless queens is that army ant colonies disperse rather poorly. Although their nomadic lifestyle can lead to genetic mixing across contiguous habitats (Soare et al. 2020), colonizing suitable terrain across any substantial body of water is essentially impossible. Army ant males, on the other hand, disperse on the wing, and *Eciton burchellii* males can probably fly for several kilometers (Berghoff et al. 2008). Studies with molecular markers have shown that this type of sex-biased dispersal is indeed reflected in the genetic structure of army ant populations. Mitochondrial markers, which are only inherited via females, can show strong separation even across rivers, but spatial structure at nuclear markers, which are inherited via both sexes, becomes apparent only at much larger geographic distances (Berghoff et al. 2008; Jaffé et al. 2009; Kronauer et al. 2011b; Pérez-Espona et al. 2012; Barth et al. 2013).

This type of information has important implications for local evolutionary processes, including how natural selection shapes the interactions between neighboring colonies (Kronauer et al. 2010), the sensitivity of army ants to habitat destruction (Pérez-Espona et al. 2012; Soare et al. 2014), and the evolution of reproductive isolation and speciation between adjacent army ant populations (Cronin et al. 2013). As we will see in the next section, genetic markers have been immensely useful in studying army ant mating biology not only at the population level but also at the level of the colony and the individual queen.

Mating Biology

The exact circumstances of army ant mating are essentially unknown, and much of the action has to be inferred from staged laboratory observations and genetic testing of the offspring. In fact, the only army ants that have ever been caught in the act in nature is a couple of *Neivamyrmex kiowapache* (formerly *Neivamyrmex carolinensis*) that were encountered when the subterranean nest was opened (Smith 1942; Kronauer and Boomsma 2007a). Luckily, once brought into the laboratory, army ant males are often not averse to a tête-à-tête with a queen, which is maybe no surprise given that their chances of finding a receptive female to begin with are probably slim at best.

FIGURE 5.10 During mating, the *Dorylus molestus* male holds the queen at the petiole with his mandibles. This copulation was staged and involves a winged male that had been collected at a light, along with an older queen that was already laying eggs. In nature, young army ant queens mate only during a brief window before they begin egg-laying, and males lose their wings before mating when entering foreign colonies. Mount Kenya, Kenya.

The army ant male initiates mating by grabbing the queen with his mandibles behind the petiole, which in *Eciton* queens is adorned with impressive horns. He then probes until he finds the tip of the queen's gaster, which he grabs with a set of claspers. During mating, males usually hold on to the queen's gaster with their legs. Army ants are persistent when it comes to lovemaking, and staged copulations lasting for ten hours have been observed in both *Eciton hamatum* and *Dorylus molestus* (Schneirla 1949, 1971; Rettenmeyer 1963a; Kronauer and Boomsma 2007a) (Figure 5.10). To what extent this reflects their natural mating behavior remains unclear, of course.

For mating to occur, the males have to locate a receptive queen in the vastness of the tropical rainforest. How exactly this feat is accomplished is not known, but it seems likely that volatile pheromones play an important role. It is therefore possible that the queen's exuberant endowment with exocrine glands is relevant here as well. Strictly speaking, however, it is unclear whether this putative pheromone emanates from the queens themselves. In the example described in the introduction to this chapter, it is questionable whether virgin queens were still present in the statary bivouac at the time the foreign males flew in, and males can also target emigration columns of conspecific colonies, rather than the young queens directly (Figure 5.11). On the other hand, it is also possible that the queen's fragrance simply lingers. In African driver ants, however, it has been reported that males can even follow abandoned raiding trails to encounter a nest, so they must use other cues in addition to putative queen pheromones (Raignier et al. 1974). Once a male has entered a foreign colony, he will shed his wings, a process that is assisted by the resident workers (Figures 5.11 and 5.12). From this point onward, in principle at least, the workers have full power over the male, in that it is up to them whether any particular male makes it through to the queen. In other words, the males have to "run the gauntlet of the workers" (Franks and Hölldobler 1987, 229), and that requires a lot of spunk indeed.

Given that the main function of ant males is to mate rather than to partake in the social life of the colony, they have been somewhat disparagingly (and, some may argue, unjustly) referred to as flying sperm

FIGURE 5.11 The moment an army ant male lands at a foreign conspecific colony, the resident workers will clip his wings. Here, an *Eciton hamatum* male has touched down next to an emigration column and is undergoing wing removal. Workers carrying brood can be seen in the background. The pruned male will then march along in the emigration, hoping for a rendezvous with the young queen.

missiles (Heinze 2016). However, for army ant males this seems like a fairly apt description. The massive African driver ant males, also known as "sausage flies," are not only the largest ant males, but each one of them carries on the order of 100 million sperm, about the same as a human ejaculate (Kronauer and Boomsma 2007a) (Figure 5.13). And even the somewhat smaller *Eciton burchellii* males still hold a respectable 10 million sperm or so (Kronauer and Boomsma 2007a). Just like the queens of other ants, army ant queens only mate during a short period early in life before they begin to lay eggs (Kronauer and Boomsma 2007a). The sperm they acquire during this period is stored in a specialized organ, the spermatheca, where the gametes are kept alive until they are used to fertilize an egg, often several years later.

However, during her lifetime, an army ant queen will likely require even more sperm than can be provided by a single male. Schneirla (1971) estimated that an *Eciton burchellii* queen lays at least 250,000 eggs per colony cycle of about thirty-six days; Franks (1989a) conservatively

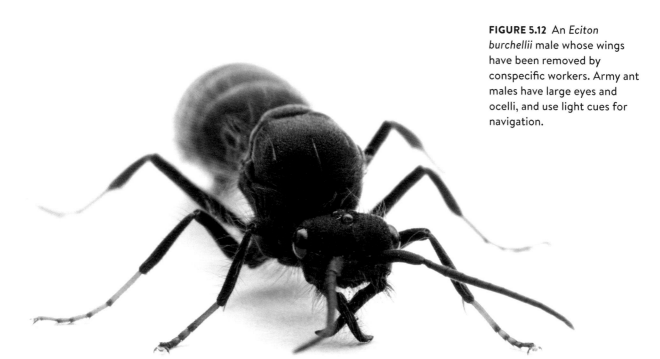

FIGURE 5.12 An *Eciton burchellii* male whose wings have been removed by conspecific workers. Army ant males have large eyes and ocelli, and use light cues for navigation.

proposed on the order of 100,000 eggs per cycle. Let's further assume that army ant queens live for about six years. Nobody really knows this, of course, but in a mark and recapture study on Barro Colorado Island, queens of both *Eciton hamatum* and *Eciton burchellii* were found healthy and well four years after they had been marked, providing a rough minimum estimate of queen life span (Schneirla and Brown 1950; Rettenmeyer 1963a; Schneirla 1971). That brings us to somewhere on the order of 6 million to 15 million eggs that an *Eciton burchellii* queen fertilizes during her lifetime, possibly more if queens live longer. In African driver ants, that number might be closer to 250 million, based on an estimate that a queen produces about 3.5 million eggs per month (Raignier and van Boven 1955).

Finally, fertilization is not perfectly efficient, in the sense that more than one sperm cell usually has to be dispensed to ensure fertilization of an egg. Accordingly, queens of the African driver ant *Dorylus molestus* store about 400 to 900 million sperm cells, and the single *Eciton burchellii* queen that has been studied had about 56 million sperm cells in her spermatheca (Kronauer and Boomsma 2007a). Given that the studied queens were of unknown age, the actual numbers in young queens right after insemination must be even higher. This implies that army ant queens mate with several males to become fully inseminated.

In most ant species, queens usually mate with a single male, and it is believed that this was also the case when eusociality first evolved in the ant lineage (Hughes, Oldroyd et al. 2008). This makes sense because, as we will discuss in more detail, in order for evolution to produce insect workers that forego their own reproduction to help a queen reproduce, the workers have to be genetically related to the sexual offspring of that queen. The stronger these genetic bonds, the more likely eusociality is to evolve, and, in sexually reproducing organisms at least, relatedness is highest in families of full siblings. As eusociality becomes more elaborate over evolutionary time and the reproductive potential of workers is more and more reduced, social evolution passes what has been called "the point of no return": once some individuals have lost the ability to mate and start their own colonies, the worker caste and thus eusociality are locked in (Wilson 1971; Boomsma and Gawne 2018). Now that the

FIGURE 5.13 The males of *Dorylus* army ants are particularly large and beefy, which has earned them the name "sausage flies" throughout Africa. *Dorylus* driver ant males measure somewhere around 30 to 35 millimeters in total body length (Raignier et al. 1974). This male of *Dorylus molestus* appeared at a streetlight near Chogoria at Mount Kenya, Kenya.

workers have no viable alternative options anymore, more promiscuous mating systems and societies with lower levels of genetic relatedness should become more likely (Hughes, Oldroyd et al. 2008).

However, even among those ant species that have passed the point of no return, most still have retained their ancestral mating system. Why? The simplest explanation may be that mating is probably costly for queens in terms of energy and time expenditure, or an elevated risk of injury and predation during the mate search and copulation (Kronauer et al. 2007a). At the same time, mating with more than one male in most cases might carry no particular benefits that would offset these costs. A handful of ant lineages, however, have curiously abandoned monogamy in favor of polyandry (from Greek, meaning "many males"), and a substantial body of scientific inquiry has been dedicated to understanding why (Boomsma et al. 2009). A first hint comes from the fact that the most prominent cases of polyandry in ants are found in lineages with very large and complex societies, such as the leaf-cutting ants of the genera *Acromyrmex* and *Atta,* harvester ants of the genus *Pogonomyr-*

mex, and, last but not least, the army ants (Boomsma et al. 2009; Hughes, Oldroyd et al. 2008).

With the advent of molecular genetic markers came the ability to examine large series of army ant workers from different colonies and to reconstruct the genotypes of both their mothers and fathers. A series of such genotyping studies have established that army ant queens are not only polyandrous but that they mate with more males than the queens of any other ants, including leaf-cutting and harvester ants. In all army ants that have been investigated in this respect, the average queen mates with at least a dozen males. These army ants include *Aenictus laeviceps, Dorylus molestus, Dorylus wilverthi, Eciton burchellii, Eciton mexicanum, Labidus praedator, Neivamyrmex nigrescens,* as well as a few other species for which the available data are more preliminary (Kronauer et al. 2004, 2006a, 2006b, 2007a; Denny et al. 2004; Jaffé et al. 2007; Barth et al. 2014; Butler et al. 2018). However, individual queens have been found to mate with as many as twenty-one males in *Aenictus laeviceps,* thirty-two males in *Dorylus molestus,* and twenty-five males in *Eciton burchellii* (Kronauer et al. 2004, 2007a; Jaffé et al. 2007). In each case, the males are unrelated to the queens they mate with, consistent with the idea that army ant males disperse over long distances. Unfortunately, the mating systems of non–army ant dorylines have not been studied so far, so we do not know whether the queens of these species adhere to the ant "standard" or whether they are army ant–like in their level of promiscuity.

However, as we have seen in Chapter 2, the colonies of these species tend to be orders of magnitude smaller than those of their army ant relatives, and the queens are much less exaggerated in their reproductive capacities. This suggests that, like most other ants, the queens of many non–army ant dorylines probably mate with only a single male. Based on what we now know about the phylogenetic relationships between the different doryline genera, this opens the exciting possibility that extreme polyandry in doryline army ants has several independent evolutionary origins, once in the New World army ants, and once or possibly even twice in the Old World army ants (see Figure 2.13). This is important, because if polyandry and the army ant adaptive syndrome had evolved in concert just once, one would have to be careful not to exclude the possibility of coincidence. However, in cases where biological traits have evolved together repeatedly, the evidence that there is some sort of functional connection between them becomes much stronger, especially if the traits are rare overall and therefore unlikely to co-occur by chance.

Recall that the first thing that makes doryline army ants stand out is the sheer size and complexity of their societies. As a very proximate reason, army ant queens have to mate with several males to get fully inseminated and to produce the large numbers of workers that are required to maintain and grow their colonies. But that does not mean that polyandry has evolved for that reason in the first place. In fact, it has been suggested that males should be able to evolve larger ejaculates if they were under selection to do so (Kronauer and Boomsma 2007a). Instead, the ultimate reasons why army ant queens have become so promiscuous might lie in the resulting increase in genetic diversity among the workers, and different hypotheses have been proposed in this context (e.g., Kronauer et al. 2004, 2007a, 2011a; Jaffé et al. 2007; Boomsma et al. 2009; Barth et al. 2014). First, the larger a society, the more likely it will be that one of its members contracts an infection of some sort, be it a parasite or a virus. Such infections are more likely to spread through and devastate genetically homogeneous populations, where all individuals will be susceptible. This is a problem well known in crop monocultures, where a given disease agent can rapidly wipe out the entire stock. Increased genetic diversity might therefore make army ant colonies more resistant to disease. The second hypothesis has to do with the extreme morphological and behavioral diversity among the female castes of some army ants, and it has been suggested that increased levels of genetic diversity might be important in producing a greater variety of offspring. What seems to contradict this hypothesis is the fact that even the queens of army ant genera with little or no worker polymorphism such as *Neivamyrmex* and *Aenictus* mate with similar numbers of males as the queens of *Eciton* and *Dorylus* (Kronauer et al. 2007a).

Hypotheses one and two would apply to any social insect with large and complex societies, but the remaining hypotheses are particularly attractive when it comes to explaining polyandry in species that reproduce by colony fission. The third hypothesis is related to the cost of producing diploid males in cases where the queen mates with a male that carries one of her sex-locus alleles. If a queen mates with a single male, either none or half of her diploid offspring will be sterile diploid males. On the other hand, if a queen mates with many males, the probability that all or even most of them carry one of her sex-locus alleles is extremely small. Although polyandrous queens are thus more likely to suffer from some small amount of diploid male load, they will hardly ever produce a lot of diploid males.

In species where queens found new colonies independently, there is intense competition between the many incipient colonies, and any level of diploid male load might convey a decisive disadvantage. In species with colony fission, on the other hand, every single one of the few offspring colonies is extremely valuable, yet large and stable from the outset. It might therefore be preferable to eliminate the risk of producing a small fraction of daughter colonies that would die from intense diploid male load, even if that comes at the cost of tolerating low levels of diploid male load in a larger proportion of colonies (Kronauer et al. 2007a). Whether army ants suffer from diploid male load in the first place remains to be determined. However, given that diploid males have been reported from many ants, including the related clonal raider ant (Kronauer et al. 2012), it seems plausible that they might.

The fourth hypothesis is conceptually related in that it ascribes a particular role of inbreeding to species with colony fission. Colony fission has been suggested to reduce the effective population size, which determines the amount of genetic variation that can be maintained in a population. Colony fission might therefore exacerbate problems related to inbreeding such as diploid male production, and polyandry might help ameliorate these effects (Barth et al. 2014). Finally, given that for every young army ant queen there are a few thousand males out there eager to land at her doorsteps, mating opportunities are plenty, and costs might be relatively low (Kronauer et al. 2007a; Boomsma et al. 2009). Whether any, some, or even all of these factors have contributed to the evolution of the unusual army ant mating system of course remains open to debate. However, a scenario in which the evolution of colony fission entailed the evolution of extremely male-biased numerical sex ratios and ultimately favored the evolution of extreme polyandry as colonies became very large seems likely, at least in the doryline army ants.

Similar to non–army ant dorylines, the information on queen mating frequencies in non-doryline army ants and other non-doryline species with army ant-like biology is still very limited. However, the little that is known suggests that the queens of many if not all of those species usually mate with a single male. The only study that has addressed this question directly looked at the mating system in the ponerine genus *Simopelta*, and found that all queens had indeed been inseminated by a single male (Kronauer et al. 2011a). In the leptanilline genus *Leptanilla*, the argument for monandry has been made based on the similar size of the male ejaculate and the female spermatheca (Gómez Durán 2016). More work, especially on ponerine army ants in the genus *Leptogenys*, is clearly required. However, given that ponerine army ants have never achieved the colony sizes and levels of social complexity seen in doryline army ants, natural selection may simply not have favored polyandry in the same way (Kronauer et al. 2011a).

While we are still pondering possible explanations, we do know that doryline army ant queens mate with many more males than the queens of any other ant species, their colonies reproduce via fission, and the sex ratio in sexual broods is highly male biased. Although this mode of reproduction is unique among the Formicidae, the carnivorous army ants have an unlikely counterpart in one of the dearest human companions: the sweet-toothed honeybee. Despite their many differences in general biology, honeybees and army ants have a few crucial things in common: their societies are exceptionally large and socially complex, and small colonies are unlikely to be viable. These factors seem to have favored the evolution of colony fission and male-biased sex ratios in both lineages, and, just as in army ants, honeybee queens often mate with several dozen males (Winston 1991; Palmer and Oldroyd 2000). This striking case of convergent evolution in honeybees and army ants lends additional credence to the idea that these three main components of the shared reproductive syndrome are in fact evolutionarily linked (Kronauer et al. 2007a).

Of note, even among the doryline army ants there is a single known, enlightening exception to queen polyandry, the North American *Neivamyrmex kiowapache*. In this species, queens typically mate with only one or two males (Kronauer and Boomsma 2007b). Despite this evolutionary reversal to low mating frequencies, the species has not sacrificed the high levels of genetic diversity observed in army ant colonies. If anything, colonies of *Neivamyrmex kiowapache* are the genetically most diverse among the army ants. Yet they achieve this diversity via an entirely different mechanism: their colonies contain several unrelated queens, sometimes more than ten, all of which reproduce simultaneously (Rettenmeyer and Watkins 1978; Kronauer and Boomsma 2007b). This is again consistent with the idea that mating for queens comes with a price tag: in cases where the benefits of genetic diversity are achieved via different means, queens are under selection to revert to low mating frequencies (Kronauer and Boomsma 2007b). Unlike polyandry, polygyny does not normally seem to evolve in response to selection for increased genetic diversity. In other words, increased genetic diversity seems to be more of a by-product of polygyny. Instead, polygyny is thought to usually represent an adaptation to harsh environmental conditions and a high risk of queen death (Rettenmeyer and Watkins 1978; Kronauer and Boomsma 2007b).

An Army's Demise

One could say that army ant colonies are immortal. They are in the sense that old queens with declining fertility can be replaced by young daughter queens during colony fission. However, when the reaper comes knocking unannounced, army ants, unlike honeybees, are unable to raise emergency queens, jeopardizing the persistence of the society (Schneirla 1971; Raignier 1972; Leroux 1979b). Colonies of *Neivamyrmex kiowapache* have solved this problem by keeping several queens around, in case one dies accidentally. The species occurs as far north as army ants go, and is regularly encountered in Kansas and Colorado, where winters are freezing and summers are dry and hot (Rettenmeyer and Watkins

1978; Kronauer and Boomsma 2007b). Compared with the permanently warm and humid tropics that most army ants call home, *Neivamyrmex kiowapache* has colonized a rather extreme environment, where unforeseen queen deaths might be of particular concern.

But even in tropical species, unanticipated queen death cannot be ruled out, even though such events are rarely witnessed. The only published observation of a naturally declining *Eciton* colony seems to be that of Schneirla and Brown (1950), who encountered a scattered *Eciton hamatum* bivouac that still contained the headless corpse of its former regent. A study of the driver ant *Dorylus nigricans* in Ivory Coast, on the other hand, found that queens die regularly and in various ways (Leroux 1979b). One fell off a liana and into a stream during an emigration, and others died without an apparent cause. Maybe most strikingly, *Dorylus nigricans* colonies are regularly attacked by subterranean army ants of the *Dorylus* subgenus *Typhlopone,* which are also able to crack mature termite mounds (Chapter 3). In six out of nine observed incidents, these epic battles led to the death of the *Dorylus nigricans* queen and the ultimate demise of the colony (Leroux 1979b).

Army ant researchers have also experimentally removed queens, and the resulting changes in colony behavior are in fact quite astounding. If an *Eciton* queen is taken from an emigration at night, the most immediate response of her colony is what Schneirla called "backtracking behavior" (Schneirla 1949, 1971; Schneirla and Brown 1950). During the next day, the ants will develop a heavily traveled column along the route of previous emigrations, seemingly in an effort to locate and retrieve the queen. Pheromone trails clearly play an important role in this behavior, and it has been suggested that *Eciton hamatum* can still detect pheromone trails that are a month old (Schneirla and Brown 1950; Blum and Portocarrero 1964).

Backtracking behavior might also serve another purpose, which is to locate and fuse with a sister colony shortly after fission—for example, if a young queen turns out to be infertile (Schneirla 1949, 1971; Schneirla and Brown 1950; Teles da Silva 1977a; Kronauer et al. 2010). Under normal circumstances, army ant colonies from the same species avoid each other relatively peacefully. On the few occasions where a collision between two swarm raids of *Eciton burchellii* has been observed, colonies either retreated to their bivouac before initiating a new raid in a different direction, or their advance was deflected to avoid the other colony (Schneirla 1949; Schneirla and Brown 1950; Swartz 1997; Willson et al. 2011). Similarly, *Dorylus molestus* colonies avoid other colonies by emigrating directly away from their nearest neighbor (Schöning et al. 2005b). Such peaceful avoidance makes sense, given that foraging in recently raided patches would be inefficient, and escalated conflict would likely be immensely costly (Swartz 1997; Schöning et al. 2005b; Willson et al. 2011). Orphaned colonies, on the other hand, do not show this avoidance behavior (Schneirla 1949, 1971; Schneirla and Brown 1950; Kronauer et al. 2010). They are ready to mingle. It turns out that this willingness to join forces with others is not limited to closely related colonies, like those headed by a mother or sister queen after fission. Essentially any nearby colony of the same species will do.

In what was probably one of my most daring efforts, Caspar Schöning and I decided to conduct a systematic study of army ant colony behavior after queen loss in the African driver ant *Dorylus molestus*. It soon became clear that removing queens from emigration columns was impractical. As we have seen in the previous chapter, driver ant colony emigrations can last for several days. Plus, due to the irritable elephants at Mount Kenya, it would have been impossible to monitor emigrations at night, making around-the-clock shift work impossible anyway. This left us with another option that was brutish, painful, but feasible in principle: snatching queens from bivouacs.

Unsurprisingly maybe, digging up a driver ant nest is not the most pleasant pastime. To the bewildered amusement of the villagers, Caspar, I, and two local field assistants, Joseph Murithi and Washington Njagi, opened a total of eighteen nests over the course of about two weeks, taking turns shoveling whenever the well-justified ant bites became unbearable. (I wore high rubber boots, and Caspar, for reasons I never fully understood, wore sneakers.) In the end, we were able to remove the queen from ten nests, and we found that the majority of the orphaned ants fused with a neighboring colony that was not directly related to them within a week (Kronauer et al. 2010).

Finally, orphaned army ant workers might have the option to raise a last brood of males before they vanish. Although their reproductive apparatus is laughable compared to that of the queen, workers of both *Eciton* and *Dorylus* possess small ovaries (Whelden 1963; Raignier 1972). Therefore, even though they do not mate, they might be able to lay unfertilized, haploid eggs that could develop into males. Genetic testing of males has been performed in two army ant species, *Dorylus molestus* and *Eciton burchellii,* and in sexual broods that were produced under normal circumstances preceding colony fission, all assayed males were sons of the resident queen (Kronauer et al. 2006b, 2007b). In the few cases where *Eciton* queens have been experimentally removed or failed to reproduce, the colonies never produced males, calling into question whether *Eciton* workers are indeed capable of producing viable eggs (Schneirla 1949; Teles da Silva 1977a). In *Dorylus,* on the other hand, the production of male larvae after queen loss has sporadically been observed (Raignier 1972; Kronauer et al. 2010). The outcome of such efforts is rather pathetic, however. In the previously mentioned study on *Dorylus molestus,* one of the queenless colonies produced a last brood of thirty-one male larvae, which is mingy compared to male broods raised by healthy colonies (Raignier et al. 1974; Leroux 1982; Kronauer et al. 2010). Furthermore, it is unclear whether these larvae would have ever made it to the adult stage, and it has in fact been reported that male larvae produced by driver ant workers do not develop far (Raignier 1972; Raignier et al. 1974). Clearly, this approach is anything but going out with a bang, and the fitness benefits are likely negligible. But what does the concept of biological fitness mean for an army ant worker anyway? Let's consider this question in the framework of inclusive fitness theory.

Inclusive Fitness Theory

To fully understand the evolution of insect societies, it is imperative to understand the concept of inclusive fitness, the idea that the biological fitness of an individual is not only determined by its own offspring (the direct fitness component), but also by the offspring of related individuals (the indirect fitness component). For example, it is impossible to account for the evolution of nonreproductive worker castes in army ants without acknowledging that their biological fitness is entirely realized as indirect fitness via the offspring of related queens and males.

Charles Darwin (1859) already had a basic understanding of this relationship, as is clearly illustrated by his further discussion of driver ant workers (or "neuters," as he calls them):

> I believe that natural selection, by acting on the fertile parents, could form a species which should regularly produce neuters, either all of large size with one form of jaw, or all of small size with jaws having a widely different structure; or lastly, and this is our climax of difficulty, one set of workers of one size and structure, and simultaneously another set of workers of a different size and structure;—a graduated series having been first formed, as in the case of the driver ant, and then the extreme forms, from being the most useful to the community, having been produced in greater and greater numbers through the natural selection of the parents which generated them; until none with an intermediate structure were produced. (241)

The concept of inclusive fitness was later mathematically formalized by the British evolutionary biologist William D. Hamilton (1964a, 1964b).

With the gradual loss of direct fitness in ant workers, natural selection has shifted more and more from the level of the individual ant to the level of the colony. This shift represents one of the major transitions in evolution, a concept that applies when selection transitions from a lower to a higher level of organismal organization (Maynard Smith and Szathmáry 1995). A remarkably similar transition, at least conceptually, is the transition from unicellular to multicellular lifeforms, where the germline and different somatic tissues would be analogous to the ant queen and the different worker castes, respectively. It is this completion of the transition to "superorganismality" that has freed up the different castes in army ants and several other eusocial insects from many of the evolutionary constraints and trade-offs faced by solitary or less social organisms, and has allowed them to evolve extremely specialized physi-ologies, morphologies, and behaviors, including worker sterility and suicidal nest defense. Clearly, these would be incompatible with a soli-tary lifestyle. Inclusive fitness theory can account for the astonishing levels of cooperation observed in army ants and other eusocial insects, but it also identifies scenarios under which evolutionary conflicts can arise between different parties. The reason is that, unlike the cells in multicellular organisms, the ants in a colony are not clonally related. This means that, at least under certain circumstances, their inclusive fitness cannot be simultaneously maximized, and from an evolutionary perspective, different parties might therefore be under selection to favor conflicting outcomes.

During colony fission, for example, one might expect workers to disagree over which larvae are raised into queens, and which of the young queens ultimately end up heading offspring colonies. The reason is that the mother queen has mated with many males, and, from an evolution-ary perspective, any given worker should prefer a full-sister queen over a half-sister queen. Given how rarely colony reproduction is observed in army ants, this potential conflict has not been studied in any detail. However, based on what we know about social insect biology, army ant workers probably cannot discriminate between full-sister and half-sis-ter queens, and this kind of nepotism seems unlikely (also see Franks and Hölldobler 1987).

The same is true for honeybees, where evidence for nepotism during queen rearing and colony fission is weak at best (e.g., Tarpy et al. 2004; Rangel et al. 2009; Sagili et al. 2018). Instead, honeybee workers appear to cooperate to rear queens of high reproductive potential, irrespective of their relatedness. Similarly, there is no clear evidence that workers act nepotistically during queen elimination, when most of the young honey-bee queens die during fatal duels (Tarpy et al. 2004; Rangel et al. 2009; Sagili et al. 2018).

Finally, even once the new queen is chosen, one might propose that a worker should preferentially accompany her during colony fission if she is a full-sister. But again, given that workers probably cannot discrimi-nate between full-sisters and half-sisters it seems unlikely that this problem would arise in practice; at least in one case it was shown that full-sisters were not overrepresented among the workers who joined a young *Eciton burchellii* queen during colony fission (Kronauer et al. 2006a).

How exactly the workforce should be divvied up between offspring colonies also constitutes a possible conflict. In cases where the mother queen is superseded and each new colony is headed by a daughter queen, theory predicts that resources should be split evenly. In these cases, there is no asymmetry between the two colonies from the perspective of the average worker. In cases where only one of the offspring colonies is headed by a daughter queen, the situation is a bit different though. From the perspective of the old mother queen, it is favorable to retain more than half the workers because her direct fitness returns from producing additional sexual offspring in the future are greater than the indirect fitness gained from having her daughter reproduce. It turns out that in this case the workers' perspective is actually similar in the sense that they are more closely related to future sexual offspring of their mother queen, which will be (half-) sisters and brothers, than that of a (half-) sister queen (Bulmer 1983; Pamilo 1991; Bourke and Franks 1995; Crozier and Pamilo 1996). The empirical data on army ants are necessarily sparse and do not allow strong conclusions, but at least anecdotally it seems that resource allocation might indeed follow this general pattern.

In African driver ants, the old mother queen is typically accompanied by the bulk of the workers and worker brood when she leaves the young queens and immature males behind during colony fission (Schneirla 1971; Raignier 1972; Raignier et al. 1974). In *Eciton burchellii,* it has been reported that in one case where the mother queen was superseded by two daughter queens, the colony split nearly exactly evenly. In a case where one of the offspring colonies retained the old queen, the resulting colony with the young queen was only about one-third of the size (Schneirla 1971). Resource allocation during colony fission in army ants might therefore mirror what is seen in honeybees, where the majority of workers accompany their mother queen when she departs in the swarm that will establish a new hive (Getz et al. 1982; Winston 1991).

Another potential conflict is that over male parentage (Hammond and Keller 2004; Ratnieks et al. 2006; Ratnieks and Wenseleers 2008). The haplodiploid sex determination system of Hymenoptera leads to an interesting relatedness asymmetry, in which females are more closely related to the sons of their full-sisters (their full-nephews) than they are to their brothers. In species where the queen only mates once and all workers in a colony are full-sisters, this creates an evolutionary conflict between the workers and the queen. From an evolutionary point of view the queen prefers her own sons, but the workers collectively prefer to raise worker sons over the sons of the queen (their brothers). This conflict shifts, however, when the queen mates with many males. In this scenario, any given worker is still most closely related to her own sons, but she is now more closely related to the sons of the queen than she is to the sons of other workers, most of which will be half-nephews. Under such a scenario, workers should be under selection to prevent each other from reproducing.

Such "worker policing" can indeed be observed in many social insects. In ants, workers with activated ovaries often are aggressed by their nestmates; in honeybees, workers eat the eggs that other workers lay. Even though the reproductive potential of army ant workers is very limited at best and worker policing has never been directly observed, the fact that all assayed males of both *Eciton burchellii* and *Dorylus molestus* that had been produced in queenright colonies were sons of the queen is at least consistent with this idea (Kronauer et al. 2006b, 2007b). However, it has to be kept in mind that worker policing can also evolve to increase efficiency at the colony level, and that in many cases this might be a more potent driver (e.g., Hammond and Keller 2004; Teseo et al. 2013). So, although it has been shown that army ant workers do not produce males as long as the queen is present, we do not know whether they show policing behavior, nor what the ultimate reasons for the evolution of such behavior would have been (Kronauer et al. 2006b, 2007b).

The phenomenon of colony fusion after queen loss in army ants, on the other hand, might represent a rare case where an altruistic behavior has evolved due to population viscosity. Even though fusing army ant colonies are often not closely related via immediate common descent, neighboring colonies are on average slightly more related to each other than two random colonies in the population, given their restricted ability to disperse. All things considered, orphaned workers might therefore be best off in terms of their inclusive fitness by serving a different queen (Kronauer et al. 2010). This interpretation, however, is far from certain, and one has to take great caution when developing adaptive explanations for observed behaviors. In this case, for example, it could just as well be that the fusion with random neighboring colonies is a nonadaptive by-product of a colony's attempt to find its own queen or to reunite with its closely related counterpart shortly after colony fission (Kronauer et al. 2010).

Their unusual mode of reproduction once again underscores the position of army ants among the pinnacles of social evolution. Although most ants abandon eusociality at least momentarily during the brief stage of solitary colony founding, the army ant queen is nothing without the colony—ever. And although she and her workers are mortal, the colony persists through time indefinitely. This has not only interesting implications for the biology of the ants themselves, as we have seen in this chapter, but potentially also for how their interactions with other organisms evolve.

In species with independent colony founding, the founding stage constitutes a severe bottleneck during which the lineage can rid itself of parasites. This especially applies to social parasites, which do not infect individual ants but live among the ants and exploit colony resources. In army ants and other species with dependent colony founding, on the other hand, social parasites will almost certainly be transmitted vertically from mother to daughter colonies. From the parasites' point of view, the massive, nomadic army ant colonies are particularly stable and resource-rich islands in the vast ocean of the tropical rainforest. Like with any decent piece of land, it is not surprising then that myriad other organisms have managed to not only crack the army ants' defense and set foot on army ant island, but also evolve astonishing adaptations to facilitate life with their ferocious hosts. These myrmecophiles, or "ant lovers," will be the subject of the final chapter.

The Traveling Circus

Ant societies are not unlike those of humans in certain respects. Some ants keep large populations of aphids for honeydew and protein, just like humans keep cattle for milk and meat. Leaf-cutting ants cultivate mono-cultures of a specialized crop fungus for food, a practice that originated many million years before human agriculture (Branstetter et al. 2017b) (see Figure 1.5). Beyond a couple of accounts of *Aenictus* and *Dorylus* occasionally tending trophobionts for honeydew (Santschi 1933; Gotwald 1995; Staab 2014), army ants have not evolved these kinds of tight-knit mutualistic interactions with other organisms. Instead, their large colonies, like human settlements, attract a plethora of commensals and parasites—the ant version of our rats, cockroaches, pigeons, and bedbugs. A potpourri of beetles, flies, bristletails, and many other arthro-pods have wiggled their way into army ant city. They have evolved ways to deceive and withstand the ants, and they live side-by-side with their fierce hosts largely unoffended while they feed on leftover food or, in some cases, the ants and their brood themselves. Many of these species show elaborate and sometimes bizarre adaptations that allow them to function as an integral part of army ant societies.

I will never forget my first army ant colony emigration and my first encounter with the army ants' menagerie of glamorous guests. It was in the summer of 2003 in Venezuela's Henri Pittier National Park, where I had come to collect samples for my studies of army ant reproductive biology. After watching the monotonous flow of ants for a few minutes, I was stunned by a shiny, drop-shaped creature shooting down the aisle like a bright red flash. Not knowing what I was looking at, it reminded me of a human cannonball in a sparkly sequin costume. I learned later that these glitter bullets were beetles of the genus *Vatesus*. As I kept staring at the ant column in amazement, I noticed additional, less obvious travelers that only revealed their full splendor when held up to a hand lens. Some were true masters of disguise, almost impossible to pick out among the stream of ants. Others were spectacular quick-change artists, easy to identify when walking freely but transforming into motionless, smooth ellipsoids at the blink of an eye when disturbed. These marvelous supporting actors add great excitement to the ants' performance. What kind of exotic creature will be next to wander out of the tropical night and into the glow of a headlamp, as the army ants' traveling circus marches by?

Ant Lovers

From the perspective of other organisms, ant colonies and their surroundings are highly attractive real estate: they are rich in resources and safe havens free of predators—excepting the ants themselves, of course. Precisely because ants are so defensive, gaining access can seem like an insurmountable challenge, yet an estimated 100,000 arthropod species in over 100 families live in specific associations with ants, at least for parts of their life cycle (Wilson 1971; Kistner 1982; Hölldobler and Wilson 1990; Elmes 1996; Thomas et al. 2005; Ivens et al. 2016; Parker 2016). These symbionts are known as myrmecophiles, from the Greek words *myrmex* for ant and *philos* for loving. Many of these associations occur outside or on the periphery of the nest, but some myrmecophiles employ clever strategies to gain entry into the colony, often by interfering with the ants' chemical or tactile communication (Wilson 1975). Those species that have evolved to live inside ant colonies, either by flying under the radar or otherwise evading attacks from the ants, are called inquilines, or ant guests (Wilson 1971; Kistner 1982; Hölldobler and Wilson 1990; Thomas et al. 2005; Kronauer and Pierce 2011). The ant colony, or the species of ant that harbors the inquiline, is referred to as the host. Some inquilines are neutral or even beneficial, but many—

the social parasites—exploit the colony at a cost to the ants. The discussion in this chapter focuses on army ant inquilines, the many species that live inside army ant colonies and travel with the ants to new bivouac sites.

Even though ant guests are taxonomically extremely diverse, certain groups are more likely to evolve myrmecophilic lifestyles than others. Mites are certainly the most diverse and numerous. Across the distributional range of *Eciton burchellii* alone, mites from about twenty-five taxonomic families have been collected from bivouacs and raiding columns, and representatives of another twenty-five or so families have been found in the ants' refuse (Rettenmeyer et al. 2011). The numbers found among other widely distributed New World army ants with large colonies are likely similarly high. At the species level, this diversity is hard to gauge. Taxonomic expertise is difficult to come by, and nobody has ventured into the biodiversity of army ant mites using DNA barcoding. A careful study of 3,146 individual *Eciton burchellii* workers from Panama found at least twenty-one mite species associated with the ants, and at least another ten species were classified as likely opportunists (Berghoff et al. 2009).

Army ant mites are not only diverse, they are also abundant. It has been estimated that a single *Eciton burchellii* colony harbors on the order of 20,000 mites, and about 5 percent of worker ants carry at least one mite (Berghoff et al. 2009)—a mitey army indeed. Most of these mites are probably phoretic, using the ants as vehicles for dispersal without harming them (Gotwald 1996; Berghoff et al. 2009; Rettenmeyer et al. 2011; Okabe 2013). Some might even be beneficial, consuming and thereby reducing the load of potentially pathogenic fungi (Berghoff et al. 2009). Others are parasitic and gorge on hemolymph, in most cases probably without leading to the ants' death (Gotwald 1996; Elzinga 1998; Rettenmeyer et al. 2011). Some species in the genus *Macrodinychus,* however, are parasitoids that ultimately kill the pupae of their host, the ponerine army ant *Leptogenys distinguenda* (Brückner et al. 2017).

Army ant mites can be extremely specialized, not only in terms of their host species but also with respect to the worker caste and even the precise body part they target (Elzinga 1978; Gotwald 1996; Rettenmeyer et al. 2011). Such species have often evolved bizarre traits to mimic the ants in order to go unnoticed. *Macrocheles rettenmeyeri* arguably takes the crown when it comes to specialized adaptations to specialized functions of specific body parts of specific ants. This species only occurs with *Eciton dulcium* and is exclusively found on soldiers and the very largest regular workers. The females attach with their mouthparts to the tarsus of the hind or middle legs, but not the front legs (Rettenmeyer 1962b) (Figure 6.1). From the ant's perspective, these little suckers might be a nuisance not only due to their appetite for hemolymph but also because they cover the ant's tarsal claws, which are required when clinging together to build the bivouac (see Figures 4.17–4.19). The hind legs of the mites, however, are hook-shaped just like the tarsal claws, and they take over this specialized function seemingly without impairing their hosts

FIGURE 6.1 The mites associated with New World army ants can be highly specialized. Among the most bizarre examples is *Macrocheles rettenmeyeri*, which attaches to the tarsi of *Eciton dulcium* soldiers and takes over the function of the ants' tarsal claws. This image shows a mite attached to the tarsus of the right middle leg. A second mite is attached to the right hind leg, but is out of focus.

in any way (Rettenmeyer 1962b). *Planodiscus* mites sit on the underside of the tibia of the ant's hind or middle legs, mimicking their host's cuticle both in terms of sculpturing and the arrangement of setae (Elzinga and Rettenmeyer 1970; Kistner 1979). *Circocylliba,* on the other hand, occupy the ant's head, thorax, gaster, or even the mandibles of soldiers. The dorsum of these mites is convex and forms a protective shield, while the ventral surface is concave, facilitating tight attachment to host surfaces (Elzinga and Rettenmeyer 1974) (Figure 6.2). Finally, mites in the genus *Larvamima* mimic army ant larvae and are probably carried by the ants from bivouac to bivouac (Elzinga 1993). These are just a few examples of mites associated with *Eciton* army ants. Mites of other species are even less studied, and many discoveries surely remain to be made.

FIGURE 6.2 Mites are common ectoparasites of insects, and a particularly large number of species are associated with New World army ants. Here, a *Circocylliba* mite (probably *Circocylliba oligochaeta*) is sitting on the sabre-shaped mandible of an *Eciton mexicanum* soldier.

At least some of the very specialized mites, like *Larvamima,* can be considered true social parasites, but mites are frequent symbionts of terrestrial arthropods in general, eusocial or not (Eickwort 1990; Okabe 2013). Accordingly, even though mites are arguably the most speciose and abundant army ant parasites, the vast majority of them parasitize individual ants, rather than exploiting the social fabric of army ant colonies. In this sense, unlike the species discussed in the remainder of this chapter, most mites are not social parasites (Parker 2016).

The organisms that have become socially parasitic myrmecophiles of ants are concentrated in three insect orders: the beetles (Coleoptera), the flies (Diptera), and the Hymenoptera, where social parasitism has evolved repeatedly among solitary wasps (Wilson 1971; Kistner 1982; Hölldobler and Wilson 1990; Parker 2016). These three orders share several features that arguably constitute broad preadaptations to the evolution of myrmecophily (Parker 2016). First, unlike many other insect orders, they lack aquatic immature stages, which would likely constrain the association with ants. Second, they are polyphagous, meaning that, across the entire group, they have a broad food spectrum. In contrast, orders that predominantly feed on plants, such as true bugs (Hemiptera) or butterflies and moths (Lepidoptera), might profit comparatively less from exploiting ant colonies. Finally, they are holometabolous, implying that different life stages are relatively unconstrained to evolve different life histories. Indeed, in most cases it is only the larva or the adult that is socially parasitic, not both (Parker 2016).

Among those three orders, the beetles harbor the greatest diversity of ant social parasites (Parker 2016). This overabundance might partly be explained by the fact that beetles, with almost 400,000 described species, simply constitute the largest insect order. However, even when accounting for the size of the order, the beetles feature disproportionately many evolutionary origins of social parasitism (Wilson 1971; Kistner 1979, 1982; Hölldobler and Wilson 1990; Parker 2016). This suggests that additional factors might be at work specifically in the Coleoptera. For once, many beetles and their grubs spend much of their life in the soil, leaf litter, or rotting wood, which also happens to be prime ant habitat. Beetles are therefore particularly likely to come into regular contact with ants. These initially coincidental interactions are of course a prerequisite for the evolution of more elaborate associations (Parker 2016). Furthermore, most beetles have a thick exoskeleton. In particular, unlike in other insects, one pair of beetle wings forms a hardened shield, the elytra, that protects the delicate wings used in flight. Their defensive armor must serve beetles well during aggressive encounters with ants, and as we will see, specialized beetle inquilines have often taken this feature to the extreme (Parker 2016). The fossil record shows that this association between ants and socially parasitic beetles is ancient (Parker and Grimaldi 2014). In fact, the evolutionary predisposition of certain beetles to become social interlopers is so strong that they were already taking advantage of ant colonies 100 million years ago in the Cretaceous, right around the time when modern ants stepped on the stage (Zhou et al. 2019).

On the ants' side of the equation, the one thing that seems to matter most in terms of myrmecophile biodiversity is colony size. The larger the mature colonies of a species, the greater the diversity of associated guests, on average (Wilson 1971; see also Päivinen et al. 2003; Campbell et al. 2013). Army ants live in colonies that rank among the very largest on the planet, and their assortment of social parasites is second to none. Why is colony size so important? First, larger colonies are more tempting targets to infiltrate because they contain more resources that myrmecophiles can feed on. However, army ants might have an edge even over other ants with very large colonies. Unlike leaf-cutting ants or seed-harvester ants, for example, army ants are predatory. During the early evolutionary stages, their carnivorous nature is probably particularly conducive to the many would-be inquilines that are themselves predators or scavengers, such as rove beetles (family Staphylinidae), clown beetles (family Histeridae), or scuttle flies (family Phoridae). Second, larger colonies provide a greater diversity of microhabitats, and therefore greater opportunity for myrmecophiles to specialize on different ecological niches (Wilson 1971).

Finally, larger colonies can sustain larger and more robust populations of a given myrmecophile species and are usually longer lived, providing stable resources across longer periods of time (Wilson 1971).

Similarly, longer-lived colonies are more likely to get colonized by a given myrmecophile at some point during their life. Taken together, these factors might increase myrmecophile diversity on a per-colony level over ecological timescales via promoting colonization and decreasing the chance of local extinction (Wilson 1971). They might also stabilize myrmecophile populations more globally, promoting speciation and decreasing the risk of extinction over evolutionary timescales, thereby leading to a net increase in species diversity. Because, as we have seen in Chapter 5, army ant colonies reproduce by colony fission, they never undergo an incipient colony founding stage during which they shed their inquilines. They are also immortal, at least in principle, thus persisting at humongous sizes indefinitely (Wilson 1971; Gotwald 1995; Berghoff et al. 2009). For these reasons, from a myrmecophile's perspective, army ants might offer even more environmental stability than other ants with large colonies.

If colony size is a deciding factor, then the richest communities of army ant myrmecophiles should be those associated with the swarm-raiding species. Due to their relative accessibility, those are also among the best studied. The numbers are indeed impressive. Over 500 species of inquilines are found with only a few species of African *Dorylus* driver ants (Kistner 1979). Similarly, an overview based on decades of work found that approximately 300 animal species depend on a single Neotropical army ant, *Eciton burchellii* (Rettenmeyer et al. 2011). This comprehensive list includes the swarm-following birds, butterflies, and flies discussed in Chapter 3. However, the list of inquilines, those species that are found in bivouac samples and raiding or emigration columns, is still extensive, featuring well over forty species of mites. Second in line in terms of inquiline biodiversity, with approximately thirty species each, are the beetles and flies. These two groups are by far the most diverse and common social parasites across the army ants. Finally, about five wasps and a couple of springtails and bristletails have been described as inquilines of *Eciton burchellii* (Rettenmeyer et al. 2011).

Although this lineup is remarkable, it requires some context. First, this list was compiled across the vast distributional range of *Eciton burchellii*. However, this species is formally subdivided into five morphologically and geographically distinct subspecies, or variants, a situation not uncommon in army ants (Borgmeier 1955; Watkins 1976). For example, the *Eciton burchellii* population at La Selva Biological Station on the Atlantic slope of Costa Rica belongs to the subspecies *Eciton burchellii foreli,* while populations from the Pacific Coast are classified as *Eciton burchellii parvispinum* (Borgmeier 1955; Watkins 1976). A deep genetic divide occurs between populations from the Pacific and Atlantic sides of Costa Rica, with no evidence for ongoing gene flow (Winston et al. 2017). These genetic data strongly suggest that the two *Eciton burchellii* subspecies are in fact separate species.

This current lack of taxonomic resolution is important to keep in mind not only when discussing the biology of this and other army ants in general, but it also implies that the comprehensive list of *Eciton burchellii* guests spans a few different host species. Because several of the inquilines are probably specific to a single host and have geographically restricted ranges, the numbers of species associated with *Eciton burchellii* colonies at any given locality will be much smaller. At the same time, army ant myrmecophiles have often been collected rather haphazardly, and their taxonomy is not well established (Rettenmeyer et al. 2011). Even among the described species, some may constitute complexes of cryptic, undescribed species, similar to the situation of their host ants (e.g., Maruyama et al. 2011). This might in turn elevate the actual number of species associated with a given host. Consequently, the numbers given here only constitute rough approximations. But what is maybe more important, they do not tell us much about the community of inquilines associated with army ants in any particular population and, therefore, the underlying ecological and evolutionary interactions.

Ten years after watching my first army ant emigration in Venezuela, Christoph von Beeren and I set out to address this gap in our knowledge by studying the community of social parasites associated with army ants at one particular site: La Selva Biological Station in Costa Rica. Not only did we want to collect inquilines systematically across as many replicate colonies as possible, we also decided to cover all six local *Eciton* species: *Eciton burchellii, Eciton dulcium, Eciton hamatum, Eciton lucanoides, Eciton mexicanum,* and *Eciton vagans.* This would allow us to study both

FIGURE 6.3 Representative social parasites of *Eciton* army ants at La Selva Biological Station in Costa Rica. *Top left:* The limuloid body shape and short, retractable appendages protect *Vatesus* beetles against ant attack. This is *Vatesus* aff. *goianus,* a species that infiltrates colonies of *Eciton dulcium* and *Eciton mexicanum. Center left: Tetradonia* beetles morphologically resemble their free-living staphylinid relatives. *Tetradonia* cf. *marginalis,* shown here, has been collected with *Eciton burchellii, Eciton hamatum, Eciton lucanoides,* and *Eciton burchellii. Bottom left:* Some army ant guests have evolved a striking morphological resemblance to their hosts. These so-called myrmecoids are prominent examples of Wasmannian mimicry. This picture shows the myrmecoid staphylinid beetle *Ecitophya gracillima,* an

associate of *Eciton hamatum*. While *Ecitophya* has an ant-like waist, it does not clearly mimic the ant petiole, as is the case in some other myrmecoids. *Top right:* The myrmecophilous staphylinid beetle *Proxenobius borgmeieri* morphologically resembles free-living staphylinids yet associates exclusively with *Eciton hamatum*. When the beetle is agitated, it recurves its abdomen, possibly to expose abdominal glands for deterrence or to increase its tactile resemblance to ants (Kistner 1979). *Center right:* The wingless, myrmecophilous scuttle fly, *Ecitophora pilosula*, has been collected from colonies of *Eciton burchellii*, *Eciton dulcium*, and *Eciton hamatum*. *Bottom right:* The bristletail *Trichatelura manni* associates with all six *Eciton* species at La Selva. Host records are from von Beeren et al. (2016a, 2016b, 2018, n.d.).

the prevalence of each inquiline species across colonies of a given host as well as its host range. To resolve taxonomic controversies and detect even cryptic species, we once again employed the DNA barcoding approach described in Chapter 3, and teamed up with different taxonomic specialists for each group of organisms.

Excluding mites, and focusing only on true inquilines—in other words, the guests that accompany army ant colonies on their emigrations—we encountered sixty-five species (von Beeren et al. n.d.) (Figure 6.3). Among this sizeable assortment were twenty-seven species of rove beetles spanning four subfamilies, seventeen species of clown beetles, six species of featherwing beetles (family Ptiliidae), and one species of water scavenger beetle (family Hydrophilidae). Also represented were eleven species of scuttle flies, one species of bristletail, and one species of millipede (von Beeren et al. n.d.). As expected, with thirty-seven species, by far the largest community of myrmecophiles was associated with *Eciton burchellii,* the species with the largest colonies. Second in line, with twenty-three inquilines, was *Eciton hamatum,* followed by the remaining four species, all of which probably have slightly smaller colonies still (von Beeren et al. n.d.). Astonishingly, even though La Selva Biological Station is arguably among the best studied tropical field sites, many species were new to science.

Among the most intriguing of the new discoveries from La Selva is the clown beetle *Nymphister kronaueri,* an exclusive associate of *Eciton mexicanum* that literally rides on the ants from bivouac to bivouac by attaching with its mandibles between the ant's petiole and postpetiole (Figure 6.4). When viewed from above while attached to an ant, the beetle so closely resembles its host's behind that it had simply gone overlooked. However, once we had fortuitously collected the first specimens, and thus knew what we were looking for, it became apparent that this beetle is not rare at all—it occurs in substantial numbers in most *Eciton mexicanum* colonies (von Beeren and Tishechkin 2017).

In other cases, DNA barcoding revealed that beetles and flies that had previously been considered a single species in fact constituted a complex of several morphologically extremely similar yet genetically distinct species (von Beeren et al. n.d.). For example, it turned out that

FIGURE 6.4 The clown beetle *Nymphister kronaueri* employs a peculiar mode of transport during the colony emigrations of its host, the army ant *Eciton mexicanum:* it attaches with its mandibles between the petiole and postpetiole of medium-sized workers. Note how the beetle retracts its head and folds its legs into cuticular grooves, offering the ants virtually no point of attack.

the taxon *Vatesus clypeatus,* a shiny red, drop-shaped beetle like the one I had observed in Venezuela, contained two rather than one species (Figure 6.5). Once the beetles had been sorted into the two species based on their DNA sequences, our taxonomist collaborator, the myrmecophile specialist Munetoshi Maruyama from the Kyushu University Museum in Japan, was able to identify a small yet consistent difference in the male copulatory organ that can be used to tell the two species apart morphologically (von Beeren et al. 2016b). Importantly, while one of the two species inhabits the colonies of *Eciton burchellii, Eciton hamatum,* and *Eciton lucanoides,* the other is found only with *Eciton vagans* (von Beeren et al. 2016b, n.d.). Using our fine genetic lens, we had gone from one size fits all to tailor-made. As we will see, the discovery of such "cryptic" species can have profound implications for

FIGURE 6.5 A staphylinid beetle, *Vatesus* cf. *clypeatus* sp. 2, running in an emigration column of *Eciton hamatum*, one of its regular host species.

our understanding of host specificity and the co-evolutionary dynamics between army ants and their guests.

Discoveries of stunning new army ant myrmecophiles over the last twenty-five years have not been limited to the genus *Eciton*. For example, a new and spectacular species of scuttle fly was found inside the colonies of *Aenictus* army ants in Malaysia. The adult females of this fly have neither legs nor wings, and their wormlike abdomens are elongated and massive compared to the rest of the body. They are so specialized that they rely on their hosts to feed them. At first glance, they resemble the ants' larvae to a T, and the ants carry the immobile flies along in emigrations, just like they do with their own young (Weissflog et al. 1995; Disney et al. 1998). This fly is so unusual that it was assigned to its own new genus, *Vestigipoda* (Disney 1996). Based on this initial discovery, researchers were able to form a search image, and four additional *Vestigipoda* species have since been described from other *Aenictus* hosts (Disney et al. 1998; Maruyama et al. 2008). In fact, these flies can be quite abundant. For example, 106 females of *Vestigipoda maschwitzi* were present in a colony of *Aenictus gracilis* (Disney et al. 1998).

The *Vestigipoda* scuttle flies are so bizarrely camouflaged that they might have long eluded army ant researchers (Weissflog et al. 1995). In other cases, simply looking carefully at previously poorly studied army ants has revealed new treasure troves of obvious yet similarly unusual inquilines. One example is the ponerine *Leptogenys distinguenda,* which has been studied in substantial detail since the initial report of its army ant lifestyle (Maschwitz et al. 1989). Among fifty studied colonies, forty-six were inhabited by *Sicariomorpha maschwitzi,* a hitherto unknown species of goblin spider (Wunderlich 1994; Witte et al. 1999, 2008; Ott et al. 2015). The spider is a kleptoparasite that feeds on the ants' prey, but given its low abundance—only about four spiders live in a typical colony—its overall impact is probably small. When the ants move, so do the spiders, following emigration columns via a combination of chemical and tactile cues (Witte et al. 1999, 2008; Ott et al. 2015). Also common in the colonies of *Leptogenys distinguenda* is a snail, *Allopeas myrmekophilos* (Janssen and Witte 2003). Rather than following the colony emigrations at a snail's pace, it produces a foam that entices the ants to carry it along

(Witte et al. 2002, 2008). Other formerly unknown guests of *Leptogenys distinguenda* include mites, springtails, bristletails, beetles, flies, and woodlice (e.g., Kistner et al. 2008; Witte et al. 2008; Disney et al. 2009b; Maruyama et al. 2010; Mendes et al. 2011; Brückner et al. 2017).

But by far the biggest unknown in the study of army ant myrmecophiles is the diversity of inquilines associated with the many strictly subterranean host species. The occasional glimpse, such as the fortuitous observation of a briefly surfacing *Cheliomyrmex morosus* emigration column and some of the associated inquilines in Panama, foreshadows what remains to be unearthed (Kistner and Berghoff 2006; Berghoff and Franks 2007; Maruyama and Disney 2008). These and other studies illustrate that army ant inquilines constitute a still poorly understood microcosm, and one of the remaining frontiers in our exploration of the world's biodiversity. With a bit of endurance, fortune, and ingenuity, students of army ant guests are poised to make wondrous discoveries for years to come.

Evading Attack

Army ant myrmecophiles have evolved many remarkable adaptations to their specialized lifestyle. While studying the colony of *Eciton burchellii* in the garden of his famous brother Fritz in Blumenau, Wilhelm Müller noticed the occasional beetle running among the ants, with some bearing a remarkable morphological resemblance to their hosts (Müller 1886). The beetles collected by Müller were later described by the Austrian Jesuit priest and entomologist Erich Wasmann as the first known guests of *Eciton* army ants (Wasmann 1887, 1895). In his initial, modest report, Wasmann correctly envisaged, "If incidental collecting from a single *Eciton* nest already yielded such favorable spoils of new and interesting staphylinids, one can without doubt expect a lot of success upon more exhaustive examination of *Eciton* nests" (1887, 403; my translation). Over the coming years, Wasmann's pioneering work indeed revealed an astounding diversity of army ant inquilines with extraordinary biology, drawing the attention of additional researchers to the study of these fascinating animals (e.g., Reichensperger 1923, 1933; Borgmeier 1930; Seevers 1965; Kistner 1979, 1982; Rettenmeyer et al. 2011). Wasmann also proposed an evolutionary hypothesis to explain the striking morphological similarity between army ants and some of their guests, a level of sophistication in ant mimicry that to this day is essentially unheard of among the inquilines of other ants (Maruyama and Parker 2017).

It was Henry Walter Bates, who made detailed observations on army ant biology during his travels in South America, as discussed in Chapter 1, who came to the realization that some perfectly palatable species have evolved to mimic noxious species, a phenomenon now known as Batesian mimicry (Bates 1862). Through Batesian mimicry a harmless species gains protection from potential predators that cannot distinguish it from the toxic species they have learned to avoid. Batesian mimicry can be interpreted as a form of exploitation because the tasty species dilutes the deterrent effect of its unpalatable model. But Bates was puzzled by cases in which two unpalatable species appeared to mimic one another. Why would this form of mimicry evolve if both species already enjoyed protection independently?

Fritz Müller was the first to put forward a plausible explanation: if, say, two unpalatable species of butterflies have a common avian predator, young birds have to learn to recognize the two species separately before they avoid them, making initial mistakes in each case. However, if the two butterfly species evolve to be indistinguishable from a bird's-eye view, young birds only have to learn a single deterrent template and make fewer mistakes overall (Müller 1879). Unlike in Batesian mimicry, both species benefit from mimicking each other, making Müllerian mimicry a case of mutualism. Butterflies in the Neotropical tribe Ithomiini, such as the army ant butterfly *Melinaea lilis imitata* (see Figure 3.32), are classic examples of species that participate in mimicry complexes as unpalatable models.

Wasmann, on the other hand, suggested that ant mimicry by inquilines serves yet another purpose: social integration into the colony (Wasmann 1925). In the case of army ants, which are essentially blind, the relevant cues are most likely tactile (Kistner 1979). In other words, when the ants touch an ant-mimicking beetle, they have a hard time telling it apart from one of their own. This third type of mimicry, adaptations that facilitate living in close association with a host by resembling it, was later coined Wasmannian mimicry by another towering student of army ant myrmecophiles, Carl Rettenmeyer (1970).

The ant-like, or myrmecoid, body plan of the beetles collected by Wilhelm Müller is the most frequently cited example of Wasmannian mimicry (Wasmann 1887, 1925). Myrmecoid beetles mimic the wasp waist, or petiole, that defines Apocrita by having a constricted abdominal base. Later abdominal segments are often expanded, resulting in a bulbous appearance that more closely approximates an ant's gaster rather than a typical beetle abdomen. Some also have enlarged antennal segments that look like an antennal scape, mimicking the signature elbowed antennae of ants. The legs are often elongated, further contributing to the ant-like gestalt. Finally, some myrmecophiles even approximate their hosts in cuticular texture (Seevers 1965; Kistner 1979; Parker 2016). Interestingly, myrmecoid beetles have almost exclusively evolved in the rove beetle subfamily Aleocharinae. What makes aleocharines special? The free-living forms in this group generally live in the same habitat as many ants, and they are predatory, small, and bear a specialized gland for chemical defense. This not only ensures that the beetles will come into frequent contact with ants, but it also predisposes aleocharines to successfully enter and exploit ant colonies. This suite of traits might explain why social parasitism is so extraordinarily common in this beetle subfamily in particular (Parker 2016; Maruyama and Parker 2017). However, myrmecophily does not necessarily beget myrmecoidy. In fact, many myrmecophilous aleocharines depart very little from their free-living counterparts, at least morphologically (Seevers 1965; Parker 2016; Maruyama and Parker 2017) (see Figure 6.3).

But aleocharines have yet another trick up their sleeves. Unlike other beetles, they have a flexible abdomen and short elytra that leave much of the abdomen exposed. This has been interpreted as a preadaptation that primes aleocharines, but not other lineages, to evolve ant-like modifications to the abdomen, setting them on the evolutionary path toward myrmecoidy (Parker 2016; Maruyama and Parker 2017). It turns out that this path is incredibly tempting. A recent phylogeny based on DNA sequence data showed that at least twelve different lineages of aleocharines have evolved myrmecoidy independently (Maruyama and Parker 2017). The overall outcome, a myrmecoid body plan, is the same in each case, but the independent lineages often differ in how exactly they implement their disguise. For example, different groups have coopted different body parts into the petiole-like constriction (Seevers 1965). Myrmecoid aleocharines are thus a textbook example of convergent evolution, illustrating that, with similar starting conditions and selective regimes, evolution can produce highly consistent results, even if the mechanistic details differ in each case (Maruyama and Parker 2017).

All these ant-mimicking aleocharines are associated with army ants or ants with army ant-like lifestyles, and, as far as we know, army ant colonies of all genera are infiltrated by myrmecoid aleocharines (Kistner 1983; Hlaváč and Janda 2009; Maruyama and Parker 2017). What's more, while each genus of army ant can host myrmecoid aleocharines from different clades, each origin of myrmecoidy gave rise to a group of aleocharines that exclusively associate with a single army ant genus (Maruyama and Parker 2017). As we will discuss in more detail, this striking host specificity might even extend to the species level. This prompts a question: Why has selection pressure imposed by army ants, but not other ants, repeatedly generated myrmecoid body plans? The answer is far from clear, and is open to speculation. One possible explanation is that the blind army ants might rely more heavily on tactile cues than other ants, and by resembling their hosts in shape and texture, the beetles are better equipped to withstand scrutiny during a pat-down (Maruyama and Parker 2017). Or should we favor an alternative hypothesis altogether?

Even though myrmecoid beetles are often cited as the classic example of Wasmannian mimicry, their status as Wasmannian mimics is in fact somewhat contentious. Some have argued that, rather than duping the ants via tactile cues, their ant-like appearance serves to evade potential visual predators that avoid the ants but would happily snack on the beetles. Were that the case, the myrmecoid beetles would merely constitute an example of Batesian mimicry (Wilson 1971; Kistner 1979, 1982; Hölldobler and Wilson 1990). This type of association is in fact known from many spiders, true bugs, and other insects that live in the vicinity of ant colonies as Batesian mimics without directly exploiting the ants (McIver and Stonedahl 1993; Cushing 2012). Myrmecoid inquiline beetles, in reality, could be a bit of both. Staphylinid beetles of the aleocharine genera *Ecitophya* and *Ecitomorpha,* for example, join the diurnal raids of *Eciton burchellii,* where they might be exposed to visual predators such as the birds at the swarm front. Accordingly, the beetles not only mimic the body shape of the ants, but also their coloration (Figure 6.6).

This becomes particularly evident when comparing beetles collected from colonies of different *Eciton burchellii* subspecies that differ in tint. Given that the ants are essentially blind, this evolutionary sensitivity to host color is almost certainly not a consequence of selection pressure that stems from the ants themselves (Kistner 1979, 1982; Hölldobler and Wilson 1990). At the same time, myrmecoid beetles are also found with army ants that do not hunt above ground and during the day, and these beetles are therefore never exposed to visual predators (Kistner 1979; Parker 2016). Furthermore, myrmecoid beetles like those of the tribe Mimecitini living with *Labidus* army ants can differ strikingly in color from their hosts, and hypothetical visual predators should have little difficulty picking them out among the ants (Parker 2016). It thus seems plausible that the body shape of myrmecoid beetles represents true Wasmannian mimicry, while their coloration, in cases where it closely matches the host, should be interpreted as Batesian mimicry (Kistner 1982; Parker 2016). Most Wasmannian mimics resemble the adult ants, but there are a few known cases of larval mimicry in army ants. These include the *Vestigipoda* flies and the *Larvamima* mites discussed previously.

Tactile cues play an important role in the social integration of many army ant guests, but the sensory world of ants relies just as heavily on chemical information. In particular, colony membership is encoded by a specific mix of different hydrocarbon molecules that are displayed on the ants' body surfaces. When two ants encounter each other, they evaluate the other's badge with their antennae. If the perceived hydrocarbon profile is sufficiently different from the colony's profile, the opponent is

FIGURE 6.6 Myrmecoid beetles are socially well integrated into their host colonies and run undisturbed in the center of raiding and emigration columns, even during heavy ant traffic. This picture shows a male *Ecitomorpha* cf. *nevermanni* (in the center) leaving the bivouac of its *Eciton burchellii* host colony on a raid. The beetle resembles the ants even in coloration, which might be an adaptation to fool visual predators like birds.

classified as an intruder and met with aggression (van Zweden and d'Ettorre 2010; Sprenger and Menzel 2020).

In principle, these cuticular hydrocarbons could also serve to expose social parasites. From the parasite's perspective, it is therefore beneficial to blend in chemically (Lenoir et al. 2001; Akino 2008). One fairly straightforward, and probably the most commonly employed way to acquire the ants' smell is to snuggle up to them. In fact, many of the guests that live in particularly close association with army ants frequently and extensively groom or rub against their hosts, and thus usually resemble them in their cuticular chemical profile (Rettenmeyer 1963b; Akre and Rettenmeyer 1966; Akre 1968; Akre and Torgerson 1968; Maruyama et al. 2009; Witte et al. 2009; von Beeren et al. 2012b, 2018) (Figure 6.7). This is reflected in a comparison between *Tetradonia* beetles, which avoid direct interactions with their *Eciton* hosts and are usually found in the periphery of raiding and emigration columns, and myrmecoid *Ecitophya* and *Ecitomorpha* beetles, which frequently groom *Eciton* workers and run in the center of columns. The cuticular hydrocarbon profiles of the myrmecoid beetles were much more similar to those of the host ants, consistent with the notion that physical contact is indeed important (von Beeren et al. 2018).

However, the level of chemical resemblance does not always correspond to the level of host incensement, at least not across different parasite species. For example, chemical mimicry of the ponerine army ant *Leptogenys distinguenda* is more precise in the bristletail *Malayatelura ponerophila* than in the goblin spider *Sicariomorpha maschwitzi*. Yet when confronted aggressively by the ants, the bristletail makes a bolt for it, while the spider is an expert animal tamer that maintains contact and calms the ants down. Probably as a result of these behavioral differences, it is the spider that is attacked less and survives longer in laboratory nests (Witte et al. 2009). Follow-up studies showed that the chemical similarity to the host decreases in both the spider and bristletail when the animals are isolated from the ants for a few days (von Beeren et al. 2011b, 2012a), supporting the idea that inquilines indeed obtain cuticular hydrocarbons, and thus their mimetic chemical profiles, directly from the hosts. When the now unfamiliar smelling spiders were reintro-

FIGURE 6.7 Many socially integrated myrmecophiles acquire their hosts' chemical profile via close physical contact. Here, a myrmecoid staphylinid beetle, *Ecitophya simulans*, is grooming an *Eciton burchellii* worker.

duced into ant colonies, they were able to deal with the ants just fine (von Beeren et al. 2012a). The ant-deprived bristletails, however, elicited significantly more aggression than their chemically well-camouflaged counterparts (von Beeren et al. 2011b). This suggests that different social parasites rely on different strategies to achieve social integration, and that chemical mimicry matters more for some inquilines than for others.

Although the cuticular chemical profiles of most army ant inquilines resemble those of the adult ants, in a few instances they are similar to those of the ant larvae instead. For example, *Vestigipoda* scuttle flies are not only convincing morphological Wasmannian mimics of *Aenictus* larvae, they also seem to smell like them, potentially making it impossible for the ants to distinguish between the imposters and their own young (Maruyama et al. 2009).

Grooming and direct physical contact with the host is an obvious and frequently used mechanism by which inquilines of various types of ants acquire compounds and thereby achieve chemical mimicry. These are therefore referred to as cases of "acquired chemical mimicry" (von Beeren et al. 2012b). At least in principle, it should also be possible for myrmecophiles to evolve the required metabolic machinery and to synthetize the ants' surface recognition molecules themselves. Such "innate chemical mimicry" (von Beeren et al. 2012b) has been suggested for a few myrmecophiles of other ants (Lenoir et al. 2012), but it remains unclear whether any of the army ant inquilines have taken this evolutionary path.

Elaborate disguise that tricks the host's senses is one way to infiltrate a society, and we can only be thankful that sophisticated humanoid rodents are not exploiting the weak spots of our own social fabric—as far as we know. Another approach is to remain visible but become unassailable. Accordingly, many army ant myrmecophiles have evolved defensive strategies to deal with their aggressive hosts. Among the most common adaptations, at least in inquiline beetles and scuttle flies, is a drop-shaped, or "limuloid" body plan (Borgmeier 1963; Seevers 1965; Kistner 1979, 1982; Brown 2017). The term "limuloid" refers to the similarly shaped horseshoe crabs in the genus *Limulus*. Along with the limuloid body plan comes a shortening of the legs and antennae, a reduction in head size, and often a particularly thick cuticle (Kistner 1979). Just like the myrmecoid morphology, this syndrome has evolved several times independently among army ant inquilines, as well as some guests of other ants. When attacked, aleocharine rove beetles in the genus *Vatesus* retract their head and legs under the massively expanded thorax, which acts as a shield (Kistner 1979) (see Figures 6.3 and 6.5). The distantly related featherwing beetle *Cephaloplectus mus* shows highly similar and functionally equivalent modifications (Figure 6.8). Given the drop-like shape and smooth surface of these beetles, the ants also have a hard time grasping and manipulating them. Different limuloid beetles associated with *Dorylus* army ants simply slip through the mandibles of their hosts (Kistner 1976, 1979).

FIGURE 6.8 An *Eciton burchellii* queen sports a featherwing beetle "hat." The ptiliid *Cephaloplectus mus* has a defensive droplike, or limuloid body form that is essentially impossible for the ants to grasp with their mandibles.

Instead of taking on a drop-like shape, inquiline clown beetles look like miniature tanks. Their bodies are compact and heavily sclerotized, and similar to limuloids, their head and appendages are retractable (Parker 2016) (see Figure 6.4). Yet another protective adaptation is found in the leaf beetle subfamily Cryptocephalinae, where the larvae construct sturdy cases out of fecal matter, and some associate with ant nests, where they feed on detritus (Agrain et al. 2015). Even though they are not common guests of army ants, one species, whose larvae follow colony emigrations, has been reported from *Dorylus* driver ants (Jolivet 1952), and I recently came across an unidentified species associated with nests of the ponerine *Megaponera analis* in Kenya (Figure 6.9).

Another adaptation to reduce your target size is to be wingless, as is exemplified by most of the midget scuttle flies that zigzag along army ant emigrations. They either are born without wings or shed their wings after entering the host colony (see Figure 6.3). While the permanent loss of wings certainly reduces the scope of ant attacks, it poses another conundrum: How do you get to the host to begin with? Among myrmecophilous scuttle flies it is often only the flightless females that exploit ant colonies, while the males do possess wings (e.g., Rettenmeyer and Akre 1968; Disney and Kistner 1990, 1998, 2003; Disney and Berghoff 2005, 2007, 2008; Disney and Rettenmeyer 2007, 2010; Disney et al.

FIGURE 6.9 The case-bearing larvae of some leaf beetles in the subfamily Cryptocephalinae are associated with ant nests. This image shows three specimens collected from the nest entrance of *Megaponera analis* at Mpala Research Centre in Kenya. Although this particular association has not been studied, the beetle larvae were observed at several different nests, suggesting some level of specificity.

2008, 2009b). In a few instances, it has been observed that the males grab the females and carry them through the air, transporting them to ant colonies during the nuptial flight (Disney 1994; Disney et al. 2008). In the limuloid *Vatesus* beetles, adults initially have fully formed wings and find army ant colonies by flight. However, the wings have a predetermined breaking point, and in beetles that are collected from colonies, the wings are truncated either because they were shed by the beetles or mutilated by the ants (Akre and Torgerson 1969; von Beeren et al. 2016b). Similarly, several species of inquiline wasps in the family Diapriidae associated with New World army ants shed their wings after entering host colonies, and in some cases their wings are removed by the ants without harm to the wasp (reviewed in Kistner 1982; Loiácono et al. 2013). Some of these wasps, like *Mimopria ecitophila,* a guest of *Eciton hamatum,* are also convincing Wasmannian mimics (Kistner 1982; Loiácono et al. 2013).

Finally, some army ant inquilines defy their hosts by more aggressive means. Aleocharine beetles in the genera *Myrmechusa, Aenictonia,* and *Anommatochara,* for example, are associated with *Dorylus* driver ants, and they prey on the ants in the vicinity of the nest. When cornered, the beetles deter the ants with chemical substances secreted from the defense gland (Koblick and Kistner 1965; Kistner 1968; Kistner 1979). In fact, many of the less integrated aleocharines associated with ants readily pepper spray their hosts (Parker 2016).

As we have seen, army ant guests have evolved several traits as specific adaptations to deal with the ants. At the same time, certain adaptations that are prominent in inquilines of other ants are rare if not absent in army ant guests. For example, some guests of formicine and myrmicine ants solicit food from their hosts, often by tapping the ants' mouthparts. The ants then regurgitate a food droplet that the inquilines consume (Wilson 1971; Kistner 1979; Hölldobler and Wilson 1990). This behavior has been studied in detail in the staphylinid beetle *Lomechusa pubicollis,* a social parasite of *Formica* and *Myrmica* ants in which both the adults and larvae are fed mouth-to-mouth by the ants (Hölldobler 1970; reviewed in Hölldobler and Wilson 1990). However, although this form of food sharing via trophallaxis is a central component of social life in some ant subfamilies and can therefore be exploited by social parasites, it is poorly documented in dorylines and ponerines (Wilson 1971; Hölldobler and Wilson 1990). Instead, most army ant guests have remained predators or scavengers, and trophallaxis with their hosts is either absent or plays a minor role at best (Akre and Rettenmeyer 1966; Kistner 1966a, 1966b, 1979; Akre 1968; Akre and Torgerson 1968, 1969; Gotwald 1995).

Some inquilines also produce chemicals that either calm down aggressive ants or induce the ants to retrieve myrmecophiles that are encountered outside the nest (Wilson 1971; Kistner 1979; Hölldobler and Wilson 1990). For example, when a *Myrmica* ant meets a *Lomechusa pubicollis* beetle while foraging, the ant initially approaches the beetle aggressively. However, once the ant starts licking the beetle's abdominal tip where the appeasement gland is located, she becomes docile. She then proceeds to lick the beetle's tufts of hairs, or trichomes, surrounding the openings of adoption glands along the beetle's sides. Bewitched by the con artist's potion, she then picks up the beetle and carries it home like a prized find (Hölldobler 1970; Hölldobler and Wilson 1990).

Such appeasement and adoption substances have not been studied in army ant social parasites. However, several species, such as the aleocharine beetle *Trichotobia gracilis,* an associate of the Asian army ant *Aenictus gracilis,* bear trichomes that might be associated with glands and resemble those of *Lomechusa pubicollis* in function (Maruyama et al. 2009; see e.g. Helava et al. 1985 for clown beetles). Given how regularly myrmecophiles in general employ these kinds of drugs to manipulate their hosts (e.g., Hojo et al. 2015), chemical analyses of exudates from army ant inquilines, along with their behavioral effects on the ants, seem like a promising avenue for future research.

Keeping Up with the Colony

While either pretending to be one of the ants or being well protected against attacks certainly helps when crashing the party, army ant nest intruders face a challenge that the guests of other ants are hardly confronted with: joining the caravan when the circus hits the road. To mites, a group that is ripe with phoretic species and ectoparasites that attach to their host, this usually comes naturally. Social parasites that are derived from free-living ancestors, on the other hand, require novel adaptations to accomplish the feat.

Springtails and many beetles, especially the tiny featherwing beetles, employ the same general trick as the mites: they hitch a ride (e.g., Wasmann 1904; Wilson et al. 1954; Kistner 1966a, 1979; Witte et al. 2008) (Figures 6.10 and 6.11). One example is *Cephaloplectus mus,* which is phoretic on host ants and prey items, both during emigrations and raids (see Figure 6.11). The beetles often run back and forth across short distances in the center of ant columns, attempting to climb onto passersby. Phoresis is also common in clown beetles; one example is the peculiar strategy of the great bareback rider *Nymphister kronaueri,* as previously described. *Nymphister kronaueri* prefers workers of intermediate size; several other clown beetles ride along on *Eciton* soldiers,

FIGURE 6.10 At the right side of this photo, a springtail (Collembola) rides on a pupal cocoon carried in a colony emigration of *Megaponera analis* at Mpala Research Centre in Kenya.

FIGURE 6.11 Two myrmecophilous featherwing beetles (*Cephaloplectus mus*) hitch a ride on a newly eclosed *Eciton burchellii* male walking in an emigration column during colony fission. These beetles can be quite abundant in *Eciton* colonies and often sit on brood or prey items carried by the ants. However, males and queens are particularly attractive to the beetles (see Figure 6.8). Small white phoretic mites are visible on the male's left middle leg and the alitrunk of the worker clinging to the male.

comfortably sitting on the underside of the ant's head (Akre 1968; Tishechkin et al. 2017; reviewed in von Beeren and Tishechkin 2017).

Some inquilines are even actively picked up and carried by the ants. These species often have grasping notches or other modifications that serve as handlebars (Parker 2016). *Trichotobia gracilis,* for example, has abdominal trichomes that the ants hold on to when carrying the beetle in emigrations (Maruyama et al. 2009). When the aleocharine beetle *Mimaenictus wilsoni*—a Wasmannian mimic of its host, the Asian army ant *Aenictus laeviceps*—arrives at an obstacle in the emigration trail that it cannot pass, it makes its predicament known to the passing ants via touching them with its mouthparts. One of the alerted ants will soon pick up the beetle by its thick first antennal segments and carry it to the new bivouac (Kistner and Jacobson 1975; Maruyama et al. 2009). Finally, diapriid wasps of the genus *Notoxopria* have hornlike projections on the thorax that the ants can conveniently clutch (Kistner and Davis 1989). It almost seems like the ants have an easily exploited grasp reflex.

If you cannot catch an army ant taxi, you have to walk—or slither—on your own. One of my most memorable army ant–related encounters occurred during fieldwork at Mount Kenya with my colleague Caspar Schöning. We had been watching an emigration of the driver ant *Dorylus molestus* for a while when, suddenly, a blind snake, possibly *Afrotyphlops lineolatus,* passed by in the middle of the column (see also Loveridge 1944; Gotwald 1982, 1995; Schöning et al. 2005b; Broadley and Wallach 2009). The snake, which was about 30 centimeters long, simply seemed enormous compared with the ants and entirely out of place, giving the scene an almost surreal appearance (Figure 6.12). The blind snakes occasionally occurring with *Dorylus* driver ants have hardly been studied, but a similar, facultative association is found between North American *Neivamyrmex* army ants and the blind snake *Leptotyphlops dulcis,* which has repeatedly been observed in raiding columns (Watkins et al. 1967b). Under controlled laboratory conditions, the snakes even follow *Neivamyrmex* pheromone trails in the absence of ants (Watkins et al. 1967b, 1972; Kroll et al. 1971). When unnerved, however, the sudden movements of the snakes elicit aggression, and the snakes respond by discharging feces and a cloacal liquid to repel the ants (Gehlbach et al.

FIGURE 6.12 Blind snakes parasitize *Neivamyrmex* army ants in North America, and *Dorylus* driver ants in Africa. This preserved specimen, possibly *Afrotyphlops lineolatus*, was collected from the emigration trail of a *Dorylus molestus* colony at Mount Kenya, Kenya.

1968; Watkins et al. 1969). When I picked up the blind snake at Mount Kenya to collect it, it coiled up, and the ants, alerted by the frantic movements, immediately attacked it, easily cutting into its flesh with their mandibles. I had pulled back the myrmecophile's cloak of invisibility.

The ability to follow army ant trails has also been suggested for amphisbaenians, fossorial, wormlike reptiles that facultatively feed on subterranean army ants (de Araújo Esteves et al. 2008). Many arthropod inquilines of African driver ants are likewise able to follow their hosts' emigration trails (Figure 6.13). David Kistner, for example, collected an impressive 4,000 plus specimens of various types of myrmecophiles from a trail over which a *Dorylus wilverthi* colony had emigrated. These guests were able to pinpoint the path of the colony, even though ant traf-

FIGURE 6.13 A small subset of myrmecophiles found toward the end of a *Dorylus molestus* column at Mount Kenya, once ant traffic had essentially ceased. Reliable species identifications are not possible based on the photographs. *Top:* A rove beetle in the genus *Anepipleuronia* or *Ocyplanus*. These beetles are well integrated, feed on booty, and sometimes are even fed by the ants (Kistner 1977). I also saw them attack and drag off army ant workers. *Bottom left:* A rove beetle in the genus *Typhloponemys*. As shown here, these beetles recurve their abdomen whenever ant density is high, which might improve ant mimicry (Kistner 1979). Species in this genus range from host specialists to host generalists (Kistner 1979). *Bottom right:* A rove beetle in the genus *Derema* or a related genus. These small beetles only walk slowly and are outrun by the ants along the columns.

fic had ceased at least fourteen hours earlier. Because African driver ants alter the substrate over which they travel by clearing the path and building tunnels, their guests might use visual or tactile cues in addition to chemical cues when following abandoned trails (Kistner 1979).

That most army ant inquilines indeed rely on chemosensory cues, possibly the trail pheromones themselves, to follow or retrace the ants' emigrations has been established in the guests of New World army ants (Rettenmeyer 1962a, 1963b; Akre and Rettenmeyer 1966, 1968; Torgerson and Akre 1969, 1970). A study that exposed a wide range of myrmecophiles to army ant scent trails in the laboratory, without the ants themselves being present, found that almost all of them followed the odorous signposts. Among others, these experiments included rove beetles of the genera *Ecitophya, Ecitomorpha, Tetradonia,* and *Vatesus,* clown beetles, featherwing beetles, millipedes, and the bristletail *Trichatelura manni* (Akre and Rettenmeyer 1968). Some myrmecophiles, such as *Ecitomorpha* and *Vatesus* beetles, even distinguished between trails of their host species and trails of other army ants (Akre and Rettenmeyer 1968). Inquiline scuttle flies, on the other hand, did not follow trails in the laboratory assay. They do, however, follow army ant trails in the field, and hundreds of flies can often still be seen hopping along long after ant traffic has ceased, suggesting that they do indeed use scent marks for orientation (Rettenmeyer and Akre 1968).

Some army ant guests employ a single strategy to keep up with their hosts. *Tetradonia* and *Vatesus* beetles, for example, always undertake the journey from bivouac to bivouac on foot, and *Nymphister kronaueri* has only ever been found riding on the ants. Other species, however, prefer to combine approaches. The clown beetle *Euxenister caroli* with its long legs often strides along, but will also climb on the ants or their cargo for a brief respite, while the featherwing beetle *Cephaloplectus mus* for the most part hitches a ride but sometimes walks for short distances (Akre 1968; von Beeren and Tishechkin 2017).

The adult army ant inquilines travel with their hosts, but their immature stages usually do not. In fact, for the vast majority of army ant guests, we do not know what their immature stages look like, let alone where they live. But here again, DNA barcoding is starting to shed some

FIGURE 6.14 The immature stages of most army ant social parasites have not been found in emigrations, and thus must develop outside the colony. *Vatesus* beetles constitute a remarkable exception. Just like the adults, the larvae of these beetles walk in emigration columns and are able to follow the ants' trail pheromone. This picture shows a larva of *Vatesus* cf. *clypeatus* sp. 2 in an *Eciton hamatum* emigration.

light. For example, using this technique, some unusual looking grubs found in the refuse deposits of *Eciton burchellii* were identified as the larvae of myrmecophilic clown beetles, and it seems likely that the immature stages of other inquilines might live in the refuse or soil around army ant bivouacs as well (Caterino and Tishechkin 2006). The only known case where immature stages regularly move with the ants are the larvae of the rove beetle genus *Vatesus,* which walk in the emigration columns of New World army ants just like the adults (Akre and Torgerson 1969; von Beeren et al. 2016b) (Figure 6.14). In fact, the lifecycle of *Vatesus* beetles is synchronized with that of their phasic hosts. Whereas females have quiescent ovaries early in the nomadic phase,

FIGURE 6.15 The bristletail *Trichatelura manni* usually walks independently in army ant emigration columns. The species occurs with a wide range of New World army ants, and is shown here with *Eciton burchellii*.

they carry large and fully developed eggs toward the end of that phase. These eggs are then laid during the early statary phase, and the *Vatesus* larvae hatch shortly before the colony becomes nomadic again, just in time to accompany their hosts on the next emigrations. As the nomadic phase proceeds, however, the larvae grow and, about halfway into the nomadic phase, disappear, probably to pupate and undergo metamorphosis in the soil (Akre and Torgerson 1969; von Beeren et al. 2016b). This means that even *Vatesus* beetles do not complete their entire lifecycle associated with the ants, and that the young adults have to find new host colonies, probably guided by olfactory cues.

Dispersal to new host colonies each generation is probably the case for the majority of army ant guests, even though a few might also be able to complete their development between subsequent emigrations (Akre and Rettenmeyer 1968; Rettenmeyer and Akre 1968; Torgerson and Akre 1969, 1970). Especially hemimetabolous inquilines, whose immature stages resemble the adults behaviorally, might not require a lot of downtime before they are ready to move. In the bristletail *Trichatelura manni,* for example, even the early instars join the ants on their emigrations (Torgerson and Akre 1969) (Figure 6.15). Nevertheless, horizontal transmission—the dispersal of parasites between unrelated colonies—is certainly the most common strategy among army ant inquilines.

Ecological and Evolutionary Interactions

The term *myrmecophile* merely denotes the fact that an organism depends on an association with ants, but it does not indicate the nature or specificity of that symbiosis. Several investigators have proposed classification systems based on a species' level of social integration and type of interaction with the ants (e.g., Wasmann 1903; Wheeler 1910; Paulian 1948; Akre and Rettenmeyer 1966; Kistner 1979; Gotwald 1995). However, the lifestyles of army ant inquilines vary in so many ways that it seems almost futile to force them into discrete categories (Mynhardt 2013). Instead, our discussion here will touch on a few general concepts and trends, illustrating them with examples that represent the range of interactions. At the same time, it is important to keep in mind that it is challenging to collect quantitative data on the relationships between army ants and their inquilines, and much of what is discussed here is therefore based on almost anecdotal observations.

First, we will explore the question of how army ant myrmecophiles affect the well-being of their host colonies. Given our current understanding of epidemiology and virulence evolution, it has been argued that the life span and size of insect societies should affect not only the diversity of parasite communities but also the evolution of parasite virulence—how much harm a parasite causes during an infection (Hughes, Pierce, and Boomsma 2008). In particular, virulence should be shaped by the extrinsic mortality of the host. If some external factor is likely to quickly kill the host, the best strategy for a parasite is to maximally exploit the host while it can, even if that ultimately leads to the host's demise. If, however, the host provides a stable and long-lived resource, parasites will be under selection to exploit the host sustainably without taking too much of a toll. Rather than having a few pathogens that can wipe out a colony, social insects with large and long-lived societies should thus host diverse communities of relatively benign and chronic parasites (Hughes, Pierce, and Boomsma 2008). As we will see, the social parasites of army ants by and large conform to this prediction.

The various inquilines of army ants fall somewhere along a continuum from being parasites that harm the ants to being commensals that gain from the association but have no discernable impact on the ants' fitness. Clear mutualists that benefit the ants are unknown among army ant inquilines. Possibly the most damaging social parasites are those that eat the ants' young. The blind snakes that parasitize African *Dorylus* and North American *Neivamyrmex* species, for example, feed on ants, both adults and brood (Punzo 1974; Kley 2001; Schöning 2008). Given the relative size of these snakes, their impact on an army ant colony might be quite significant. At the same time, they are not obligate associates of army ants. *Leptotyphlops dulcis,* for example, consumes all kinds of other invertebrates as well (Punzo 1974). Blind snakes are therefore rather unusual examples of army ant inquilines, in that they infiltrate colonies only occasionally, but when they do, their impact is probably sizeable.

Among the obligate social parasites of *Eciton* army ants, *Vatesus* beetles are probably among the most virulent. Both adults and larvae will take ant brood and prey, but *Vatesus* larvae in particular voraciously consume live army ant brood (Akre and Torgerson 1969; von Beeren et al. 2016b). In a laboratory experiment, four *Vatesus* larvae ate about fifty small *Eciton hamatum* larvae over the course of several days (Akre and Torgerson 1969). *Vatesus* beetles are also abundant. In our study at La Selva Biological Station, every colony of *Eciton burchellii* and *Eciton hamatum* was infected, containing up to 130 and ninety-six *Vatesus* larvae, respectively (von Beeren et al. 2016b). If we take the extreme case of 130 *Vatesus* larvae and assume that each consumes thirty army ant larvae before leaving the colony, they would collectively kill 3,900 *Eciton* larvae during a colony cycle. This is a large number, but it constitutes less than 2 percent of the larvae in an *Eciton burchellii* worker brood (see Chapter 4). Furthermore, it has to be kept in mind that most of these larvae would be eaten at an early developmental stage before the colony has invested a lot of resources. Other social parasites, such as the clown beetles *Euxenister caroli* and *Euxenister wheeleri,* also feed on the brood and booty of their *Eciton* hosts (Akre 1968). However, the larvae of these species do not develop inside the ant colony, and the number of adults per colony is usually low. For example, Akre (1968) reported only one to nine *Euxenister caroli* specimens per *Eciton burchellii* colony, and one to sixteen *Euxenister wheeleri* per *Eciton hamatum* colony.

On the other side of the spectrum are commensals that mostly feed on detritus inside the bivouac, in the refuse deposits, or the soil around the bivouac. The bristletail *Trichatelura manni,* for example, feeds on small particles on the ants' cuticle and, less frequently, on fluid oozing from prey items. There is no evidence that the bristletail is detrimental to the ants in any way, and it has even been suggested that it fulfills some sort of hygienic function in the colony (Rettenmeyer 1963b). Similarly, millipedes of the genera *Calymmodesmus* and *Rettenmeyeria,* which are common associates of *Labidus* and *Nomamyrmex* colonies, feed on organic debris in the soil around the nest, possibly helping prevent the growth of molds (Rettenmeyer 1962a).

Given the great diversity of army ant inquilines, one wonders what their collective effect on colonies might be. Kistner (1979), for example, collected four flies and 4,557 staphylinid beetles representing more than fifty species from a single *Dorylus* driver ant emigration in Africa, and reported an outstanding incidence where he sampled 13,000 myrmecophiles from another driver ant colony. These numbers are truly astonishing. It has to be kept in mind, however, that the societies of driver ants measure many million individuals. Doing a back-of-the-envelope calculation, Kistner (1979) estimated that for just over every

5,000 ants one could expect to find one social parasite. In comparison, for approximately every four humans, one rat roams the streets and sewers of New York City (Auerbach 2014). Consequently, while the rats' insolence and mammoth proportions feature in every other New York standup comedy routine, the army ants might hardly take notice of their social parasites. In conclusion, the average number of social parasites relative to the number of army ants is quite low, suggesting that in most cases their effect on the colony is not detrimental. Unfortunately, measuring the virulence of army ant social parasites directly is essentially impossible because this would require studies in which entire colonies are experimentally infected and then compared with uninfected controls under standardized conditions.

Another important aspect of parasite biology is the level of host specificity, and it turns out that different army ant social parasites indeed have strikingly different host ranges. At La Selva Biological Station, for example, the bristletail *Trichatelura manni* and the featherwing beetle *Cephaloplectus mus* can be found with all six local *Eciton* species, while the clown beetles *Nymphister kronaueri, Euxenister caroli,* and *Euxenister wheeleri* infect only a single host species each (von Beeren et al. n.d.). The level of host-specificity is believed to be related to several other components of a parasite's biology, because parasites should face an important life-history trade-off. On the one hand, a parasite could employ a generalist strategy to infect a broad range of hosts and exploit them with reasonable efficiency. On the other hand, a parasite could evolve sophisticated adaptations that increase its efficiency when infecting a particular host species. However, such adaptations would likely come at the cost of decreasing its performance when infecting other hosts, thus limiting the parasite's host range and therefore the number of available targets (e.g., Price 1980; Schmid-Hempel 2011). In the context of army ant social parasites, one might thus surmise that morphologically and behaviorally specialized species should have narrower host ranges, on average (Ivens et al. 2016). Although this seems to be true in some cases, it is probably not a universal trend.

Possibly the best-studied comparison between "generalized" and "specialized" army ant inquilines is within staphylinid beetles (Akre and

Rettenmeyer 1966; von Beeren et al. 2018). Representative examples of generalized staphylinids are those in the genus *Tetradonia,* discussed earlier. *Tetradonia* beetles do not have obvious morphological adaptations to their inquiline way of life (see Figure 6.3). They either run in the periphery of emigration columns or during times when ant traffic is minimal—for example, at the very end of colony emigrations when the brood and most ants have passed (Figures 4.9 and 6.16). They attack and prey on army ant workers, and can frequently be seen dragging injured or weak ants away from raiding and emigration columns and into the leaf litter (Figure 6.17). They never groom the ants, their cuticular hydrocarbon profiles resemble those of their hosts only weakly, and in laboratory nests

FIGURE 6.16 Staphylinid beetles of the genus *Tetradonia* are poorly integrated guests of New World army ants that walk in the periphery of raiding and emigration columns and avoid the ants. They are also commonly observed at the very end of emigrations, when ant traffic has ceased. This picture shows two *Tetradonia* beetles running after their *Eciton hamatum* host colony. The beetle on the left cannot be identified with certainty; the beetle on the right is *Tetradonia laselvensis.* We discovered this species during an extensive survey of *Eciton* army ant inquilines at La Selva Biological Station in Costa Rica, and it is named after the type locality (von Beeren et al. 2016a).

FIGURE 6.17 Staphylinid beetles in the genus *Tetradonia* are specialized predators of New World army ants. In most cases, however, they attack injured or dying ants. This *Eciton burchellii* worker was decapitated during an altercation with an *Atta cephalotes* leaf-cutting ant soldier, before the *Tetradonia* beetle dragged off its corpse.

they hide or are otherwise attacked by the ants. *Ecitomorpha* and *Ecitophya,* on the other hand, are morphologically derived myrmecoid Wasmannian mimics (see Figure 6.3). They run in the center of emigration and raiding columns during the height of ant traffic (see Figure 6.6). They are kleptoparasites that feed on the ants' prey but do not attack adult ants, and they frequently groom workers (see Figure 6.7). Accordingly, their chemical surface profile resembles that of the ants, and the ants do not behave aggressively toward the beetles.

In summary, *Tetradonia* beetles resemble ancestral, free-living staphylinids and are poorly integrated into their host colonies, while *Ecitomorpha* and *Ecitophya* beetles show sophisticated morphological and behavioral adaptations and blend in extremely well (Akre and Rettenmeyer 1966; von Beeren et al. 2018). Broadly speaking, this difference is indeed reflected in their host range. At La Selva Biological Station, for example, *Tetradonia laticeps* occurs with all six *Eciton* species, and each of the four remaining *Tetradonia* species is likewise found with more than one host (von Beeren et al. 2016a, 2018). By contrast, both *Ecitomorpha* and one *Ecitophya* species are only found with *Eciton burchellii,* and *Eciton hamatum* is the single host of the second *Ecitophya* species (von Beeren et al. 2018). In fact, this extreme level of host specificity, combined with high levels of social integration, is characteristic for the various lineages of army ant–associated myrmecoid aleocharines (Seevers 1965; Kistner 1966a; Kistner and Jacobson 1990; Maruyama and Parker 2017; Pérez-Espona et al. 2018).

Although this example fits well with the idea of a fundamental life-history trade-off in army ant social parasites, other cases are more difficult to reconcile. For example, the rove beetle *Proxenobius borgmeieri* only occurs with *Eciton hamatum* yet has retained a generalized staphylinid body plan (von Beeren et al. n.d.) (Figures 6.3 and 6.18). Similarly, even closely related species in the scuttle fly genus *Ecitophora* can have widely different host ranges, but differences in their level of social integration are not obvious (von Beeren et al. n.d.). Among aleocharines associated with *Dorylus* army ants in Africa, some are strictly host specific, while others occur regularly with two, three, or even more host species. Among these, the level of morphological specialization does not correlate with the level of host specificity (Kistner 1979). However, the biology of these latter species has been studied less than that of the *Tetradonia, Ecitomorpha,* and *Ecitophya* rove beetles discussed earlier, and it is thus possible that similar patterns will emerge as additional data on their taxonomic status and life histories become available.

Finally, the antagonistic evolutionary dynamics between parasites and their hosts should lead to arms races during which parasite adaptations that increase virulence are recurrently met with counteradaptations of the host that reduce virulence (Schmid-Hempel 2011). In evolutionary biology, this is known as the Red Queen effect, named after Lewis Carroll's *Through the Looking-Glass* (1871), where the Red Queen tells Alice, "it takes all the running you can do, to keep in the same place" (42). Assuming that these dynamics also apply to inquiline social parasites, one might therefore expect that ants have evolved more definitive defense strategies against more harmful parasites (von Beeren et al.

FIGURE 6.18 *Proxenobius borgmeieri* is highly host specific, occurring only with the army ant *Eciton hamatum.* Here, a beetle is running in the nightly emigration column of its host colony.

2011b). At least superficially, there seems to be some evidence for this. Army ants usually ignore myrmecophilous bristletails and millipedes, for example, while they constantly chase *Vatesus* beetles in laboratory nests, forcing them into hiding (Rettenmeyer 1962a, 1963b; Akre and Torgerson 1969). Similarly, as noted previously, generalized staphylinids such as *Tetradonia* are often attacked by the ants whereas specialized staphylinids such as *Ecitophya* and *Ecitomorpha* are not (Akre and Rettenmeyer 1966; von Beeren et al. 2018). The same pattern is found among staphylinid beetles associated with Asian *Leptogenys* army ants. The ants behave aggressively toward beetles that prey on their larvae, but they are far less vigilant when it comes to nonpredatory species (von Beeren et al. 2011a; see also Parmentier et al. 2016). However, not only is it challenging to quantify parasite virulence and the level of host defense, it is also not necessarily clear what the appropriate comparisons should be, given that any two army ant inquilines can belong to closely related or highly divergent evolutionary lineages. A comprehensive and robust phylogeny of army ant social parasites would therefore be an important step toward measuring correlations between different parasite and host traits while statistically controlling for phylogenetic relationships.

As we have seen, the interactions between army ants and their guests are manifold. Each species of army ant is host to an entire community of interlopers, and some myrmecophiles are host specialists while others also infiltrate the colonies of additional species. As a consequence, the army ants in a given community are ecologically connected not only via niche overlap and separation in their prey spectra (Chapter 3) but also via a network of inquiline social parasites. The ecological and evolutionary interactions between army ants and their guests are therefore best understood in a community context, rather than as many separate one-to-one associations (Ivens et al. 2016).

In fact, it is not unlikely that interactions between different parasites contribute to shaping the local assemblage associated with a given army ant colony or species. Some parasite species might locally compete with or even exclude each other, while some might benefit from the presence of others. Diapriid wasps, for example, are often endoparasitoids of flies; even though the biology of socially parasitic diapriids associated with army ants has not been studied in detail, it has been suggested that at least some might in fact target the many myrmecophilic phorids inhabiting the same colonies (Loiácono et al. 2013). And while most of the phorids are thought to be scavengers that lay their eggs in food leftovers, phorid larvae have also been recovered as endoparasites of staphylinid and histerid beetle larvae collected from the refuse deposit (Bruch 1923).

Furthermore, the composition of the local army ant community might partly determine which social parasites are present and associated with each species of army ant. For example, a particular species might have a preferred host and infect colonies of other army ants only in communities where the preferred host is absent. As is the case in the beetle *Lomechusa pubicollis,* discussed previously, a species might depend on the presence of two different hosts to complete its lifecycle. Or species might adjust their behavior depending on the number of suitable host species in the community. None of these possibilities have been studied in army ants and their inquilines. However, they highlight the fact that

FIGURE 6.19 While inquiline social parasites live inside army ant colonies and accompany the ants on emigrations, the ants' garbage heaps, or refuse deposits, which form beneath the bivouacs, constitute a valuable resource for many addition-al, more loosely associated species. Here, a staphylinid beetle of the genus *Ecitopo-ra* is scavenging among an *Eciton burchellii* refuse sample photographed in the laboratory. *Ecitopora* beetles are frequent inhabitants of *Eciton* refuse piles, but they do not usually enter the colony or interact with the ants (e.g., Kistner and Mooney 2011).

ecological communities are networks, tightly interwoven across time and space, rather than laundry lists of species existing in isolation.

The communities centered on army ants go even beyond the many inquilines discussed in this chapter. For example, many additional arthropods live as facultative associates in the refuse deposits of army ant colonies, where they feed on deceased ants and discarded food leftovers (Rettenmeyer 1963a; Kistner 1979, 1982; Rettenmeyer et al. 2011) (Figure 6.19). Furthermore, a suite of organisms infect individual army ants as endoparasites, including fungi, nematodes, strepsipterans, flies, and wasps (reviewed in Quevillon and Hughes 2018; see also Poinar et al. 2006). Fungi can also be ectoparasites of army ants (e.g., Haelewaters et al. 2017). We still know next to nothing about the diversity and biology of these organisms, and, like with the inquilines, much remains to be explored.

Epilogue

The beautiful illustrations of Maria Sibylla Merian, including her plate depicting army ants on a guava tree, convey a deep appreciation for the interconnectedness of all lifeforms (see Figure 1.1). At a time when most naturalists were merely interested in cataloging species, Merian anticipated Charles Darwin's famous metaphor of life on Earth as a "tangled bank" by a good 150 years.

Army ants epitomize the tangled bank. Chapter 3 discussed their role as top arthropod predators, and how entire guilds of birds and flies depend on their majestic swarm raids. In Chapter 6, we have seen how army ant communities interact with highly diverse networks of social parasites. Yet the ecological entanglements of army ants are even more varied. The many subterranean species move sizeable quantities of earth, which undoubtedly affects soil aeration at an ecosystem level (Gotwald 1995; Schöning 2008). For example, a *Dorylus* driver ant colony excavates up to 20 kilograms per day while digging a new nest cavity (Leroux 1977b). Army ants also constitute an important food source for other organisms (e.g., Leroux 1979b; de Souza and Calouro 2018; reviewed in Gotwald 1982, 1995). Maybe most famously, chimpanzees feed on *Dorylus* driver ants by inserting long sticks into the nest. Once the disturbed ants ascend the wand, the chimpanzee withdraws the stick and pulls it through its mouth, a behavior called "ant dipping" (reviewed in Gotwald 1982, 1995; for more recent studies see, e.g., Sugiyama 1995; Schöning et al. 2007, 2008b; Sanz et al. 2010; Koops et al. 2015a, 2015b). Ants are proverbially small, but their ecological impact is simply enormous.

Unfortunately, the insatiable and ruthless human disposition has become a seemingly insurmountable threat everywhere on this planet. Especially the mindless destruction of tropical rainforests around the world is endangering army ant populations and wiping out the associated biodiversity. Many of the swarm raiding species are sensitive to habitat degradation and will soon go extinct in small forest fragments and cleared areas (Lovejoy et al. 1986; Harper 1989; Partridge et al. 1996; Roberts et al. 2000b; Meisel 2006; Schöning et al. 2006; Kumar and O'Donnell 2009; Peters et al. 2009, 2011; Baudier et al. 2015). In the longer term, species like *Eciton burchellii* might go extinct even in larger and only partially fragmented reserves (Boswell et al. 1998; Britton et al. 2001).

With the disappearance of army ants undoubtedly comes a cascade of coextinctions. Needless to say, their demise will go hand in hand with the loss of the many obligate social parasites. However, given the vast ecological reach of army ants, many additional species will be affected in ways that are more difficult to predict, with the impoverishment of prey communities being one likely outcome. The ant-following birds are the only ones studied explicitly in this respect; indeed, both in the African and American tropics, specialist birds vanish from forest fragments too small to sustain populations of swarm raiding army ants (Harper 1989; Stouffer and Bierregaard 1995; Stouffer et al. 2006; Kumar and O'Donnell 2007; Peters et al. 2008; Peters and Okalo 2009; Ocampo-Ariza et al. 2019).

The more we destroy the natural world around us, the less we are exposed to its marvels. The less we become acquainted with nature, the less we care about it, and the more we are willing to destroy it further. It becomes a vicious cycle. Over the last twenty-five years, we have come to a point where immediate and drastic adjustments to how we value and treat nature and our planet are imperative if we want to save what is left. This might sound daunting, but it is not impossible. Yet it will require environmental awareness in each one of us, and determined political action guided by scientific rationale to achieve long-term sustainability.

Another recurring theme throughout this book has been the profound effect of group size on social organization. Along the early evolutionary trajectory of army ants, scout-initiated group raiding turned into spontaneous army ant mass raiding as colony size gradually increased (Chapter 2). Later, some army ants shifted from column raiding to swarm raiding with further increases in colony size, accompanied by concurrent dietary expansions (Chapter 3).

In Chapters 2 and 5, I have touched on the issue of caste development and evolution, and how the level of caste polymorphism also tends to increase with colony size. As the dichotomy between the queen and worker caste became more and more pronounced, some lineages approached an important transition in eusocial evolution. At this point of no return, natural selection had shifted almost entirely to the level of the colony, eliminating evolutionary developmental constraints related to worker reproduction. Workers were now free to evolve extremely specialized morphologies. This trend reached its pinnacle in some of the army ants with particularly large colonies, massive queens, and highly polymorphic workers.

This transition also relaxed constraints on the mating system, because high genetic relatedness among colony members was no longer a strict requirement for social cohesion. Accordingly, in some of the largest insect societies, queens evolved to mate with many males. The two clades of doryline army ants in particular stand out in their extreme levels of polyandry (Chapter 5).

Finally, in Chapter 6 we have seen that the larger an ant colony, the more attractive it is to social parasites. Especially the swarm raiding army ants arguably host more diverse communities of associated guests than any other social insects. The tale of the army ants truly is one of superlatives.

Throughout this book, I have emphasized how DNA sequencing has transformed our understanding of these and other topics over the past two decades. Based on molecular data, we now have a clear understanding of the ant family tree and the phylogenetic relationships between the different army ant genera and their relatives, enabling us to reconstruct different aspects of army ant evolution with greater confidence than ever before (Chapter 2). DNA barcoding allows us to reliably identify prey items, and to finally study the prey spectrum of different army ants at the species level (Chapter 3). The same technique has been used to delimitate species boundaries and reconstruct life cycles of army ant social parasites, providing unprecedented insights into the evolutionary and ecological dynamics underlying host–parasite interactions (Chapter 6). Molecular studies have also revealed the unusual mating and reproductive system of army ants, and how it is reflected in the genetic structure of populations (Chapter 5).

Although most of this work was based on a few short stretches of DNA, researchers can now sequence entire genomes, the complete genetic information of an organism, at comparatively low cost. Taking advantage of this, we have produced whole genome sequences of the first doryline, the clonal raider ant *Ooceraea biroi* (Oxley et al. 2014; McKenzie and Kronauer 2018). This resource now allows us to identify candidate genes that might underlie and regulate anything from pheromone communication to phasic colony cycles (e.g., McKenzie et al. 2016; Libbrecht et al. 2018; Chandra et al. 2018).

Two additional recent technological breakthroughs have enabled strides in the study of ant biology. First, it is now possible not only to describe the genomic sequence of organisms, but also to edit it. This approach, which we recently pioneered in the clonal raider ant, is indispensable for understanding the function of candidate genes (Trible et al. 2017). Second, advances in machine learning now allow us to automate the high-throughput collection of behavioral data, another technique we have established in the clonal raider ant (Ulrich et al. 2018, 2020; Gal et al. 2020; Chandra et al. n.d.).

In concert, these powerful techniques can provide many novel insights into doryline biology, using the clonal raider ant as a laboratory model system. Although army ants cannot be maintained in captivity for extended periods and many of these techniques will therefore not be applicable, what we learn from the clonal raider ant will likely inform targeted molecular studies that can in fact be carried out in the field. As a first step in this direction, we have recently sequenced the first army ant genome, that of the swarm raider *Eciton burchellii* (McKenzie et al. n.d.). What we will learn over the next twenty-five years is of course impossible to predict. But that is the beauty of science: there is something unexpected and potentially enlightening behind every corner. My cautious anticipation is that whoever ventures into the world of army ants with new tools at hand will return with many new fabulous stories to tell.

This book is my contribution to documenting the beauty of the natural world and making it accessible to others. I hope that it will help you appreciate army ants just a little bit more and recognize the urgent necessity of preserving nature in all its complexity as a central part of our cultural heritage. Luckily, unlike the ants, we can use rational thinking and foresight to steer our collective behavior and environmental impact, as our own societies grow ever larger and more integrated.

Maybe this book will even inspire you to go out and explore. There is so much left that we don't understand and so much that remains to be discovered. I have pointed out a few open questions throughout the book, but there are many more worth pursuing. Especially with the advent of new technologies in genetics, neuroscience, and ethology, we live in an exciting era where scientific problems that seemed entirely out of reach just a few years ago are now becoming accessible. Identifying the most relevant questions and properly putting results into context, however, will always require a deep understanding of basic biology. It is this amalgamation of thorough natural history with new experimental approaches that holds the greatest promise to reveal some of the world's remaining wonders.

glossary

Aculeata: A **clade** in the insect order **Hymenoptera**, also called the "stinging wasps." The defining feature of the Aculeata is that the ovipositor has been modified into a stinger. They include the ants (family **Formicidae**), bees, **eusocial** wasps, and several groups of **parasitic** wasps.

Age polyethism: Age-related behavioral differences and division of labor. In **eusocial** insects, young workers often tend to perform tasks inside the nest, such as nursing, while older workers perform tasks outside the nest, such as foraging and colony defense.

Alate: Bearing wings; an individual with wings. From the Latin word *ala* for wing. In many ants, **queens** and males are born with wings. In ants, alate **queens** lose their wings after mating and are then referred to as **dealate**.

Alitrunk: The middle part of an ant's body, between the head and the **petiole** (see Figure P.6).

Allele: A variant at a particular locus in the genome.

Allometry: The study of how different body parts change relative to each other as the overall body size of an organism changes.

Altruism: A social behavior that, on average, conveys a lifetime direct fitness benefit to the recipient but comes at a lifetime direct fitness cost to the actor; see **inclusive fitness**.

Antenna (plural antennae): Paired sensory appendages on the head of an **arthropod**; colloquially referred to as the "feelers" (see Figure P.6).

Antennal scape: The basal segment of the **antenna** (see Figure P.6).

Apocrita: A suborder within the insect order **Hymenoptera**. Apocrita are defined by the presence of a "wasp waist," or **petiole**. They are composed of the **parasitoid** wasps and the **Aculeata**.

Army ant: An ant **species** that shows all components of the **army ant adaptive syndrome**. The vast majority of army ants are found in the ant **subfamily Dorylinae**.

Army ant adaptive syndrome: The army ant adaptive syndrome consists of three functionally and evolutionarily interrelated traits: spontaneous **mass raiding**, **nomadism**, and **colony fission**.

Arthropod: A member of the phylum Euarthropoda, which includes arachnids, crustaceans, insects, and myriapods.

Bivouac: The temporary **nest** of an **army ant colony**; a cluster made up entirely of the bodies of live **workers**, housing the **queen** and **brood**.

Brachypterous: Having highly reduced, usually nonfunctional wings.

Brood: Collectively the immature stages present in an ant **colony**: eggs, **larvae**, and **pupae**.

Brood care phase: See **nomadic phase**.

Brood-stimulative theory: Theodore Schneirla proposed that different **brood** developmental stages regulate the **colony** cycles of **phasic army ants**.

Budding: See **colony budding**.

Callow: A young adult ant shortly after **eclosion**, whose **cuticle** is not yet fully melanized (see **melanin**).

Caste: In ants, a set of females that is morphologically distinct. The term is sometimes extended to behaviorally distinct females, even in the absence of morphological differences.

Clade: An evolutionary entity that contains a common ancestor and all of its descendants.

Cocoon: A pupal casing spun by the **larva** out of silk. In **army ants**, some **genera** such as *Eciton* and *Labidus* spin cocoons, and others such as *Dorylus* and *Aenictus* do not. In *Neivamyrmex,* **queen** and male **larvae** spin cocoons, but **worker larvae** do not.

Collective behavior: The coordinated behavior of individuals in large groups, usually via local interaction rules, and its emergent properties. Commonly cited examples are flocking in birds, schooling in fish, and **nest** construction in social insects.

Colony: A group of individuals of the same species, more than just a mated couple, that communicate to construct **nests** and raise offspring cooperatively.

Colony budding: A type of **dependent colony founding** in which **polygynous colonies** bud off **colony** fragments containing both **queens** and **workers**. Offspring **colonies** usually remain close to their mother **colonies**.

Colony fission: A type of **dependent colony founding** employed by **army ants** and honeybees. When **colonies** of **monogynous** species split in two, each resulting **colony** is again headed by a single **queen**. Offspring **colonies** usually do not remain close to their mother **colonies**.

Column raid: A type of **mass raid** in which **army ants** leave the **nest** in a column. "Pushing parties" of ants can fan out into miniature swarms at the raid front. As opposed to **swarm raids**, column raids typically occur in **army ants** with more moderate **colony** sizes (see Figure 3.13).

Commensal: See **commensalism**.

Commensalism: A type of **symbiosis** between two **species** from which one partner benefits while the other has neither a **fitness** gain nor loss.

Cooperation: A social behavior that, on average, increases the lifetime direct fitness of all parties involved; see **inclusive fitness**.

Cretaceous: A geological period extending from ca. 145 to 66 million years ago.

Cuticle: The structure forming the **arthropod exoskeleton**.

Cuticular hydrocarbons: Organic compounds on the **cuticle** of **arthropods** that impede desiccation and can also function as **pheromones** in ants and other insects.

Dealate: A formerly **alate** insect that has lost its wings.

Dependent colony founding: Queens are accompanied by **nestmate workers** during **colony** founding.

Dichthadiiform: Referring to a **dichthadiigyne**.

Dichthadiigyne: A **queen** with an enlarged **gaster**, capable of extreme **physogastry** and a high reproductive output. Dichthadiigynes are also **ergatoid**.

Diploid: Having two complements of chromosomes, one usually being inherited from the mother, the other from the father.

Dorylinae: The ant **subfamily** that contains the vast majority of **army ants** as well as many additional **species** with army ant-like behavior.

Doryline: Pertaining to the ant **subfamily Dorylinae**.

Driver ant: An **army ant species** that drives fleeing **arthropods** in front of its **epigaeic swarm raids**. Historically, the term has mostly been used for a **clade** of **species** in the African *Dorylus* subgenus *Anomma,* and this is how it is used in this book. However, the same phenomenon occurs in the **Neotropical species** *Eciton burchellii* and *Labidus praedator.*

Eclosion: The act of the **imago**, or adult, emerging from the **pupa** at the end of **metamorphosis**.

Ectoparasite: A **parasite** that lives outside and on the surface of the **host**'s body.

Emigration: A **colony** relocation; frequent **colony** relocations are the defining feature of **nomadism**.

Endoparasite: A **parasite** that lives inside the **host**'s body.

Endoparasitoid: A **parasitoid** that lives inside the **host**'s body.

Entomology: The scientific study of insects. The term is derived from the Greek words *entomon* for insect, and *logos* for study.

Eocene: A geological epoch extending from ca. 56 to 40 million years ago.

Epigaeic: Active above ground; most **army ants** are **hypogaeic**, but some conduct epigaeic **mass raids** or **emigrations**, and a few even construct epigaeic **bivouacs**.

Ergatoid: Literally meaning "**worker**-like" but specifically referring to a permanent absence of wings in the **queen caste**.

Eusocial: See **eusociality**.

Eusociality: An extreme form of social living, which is defined by three traits: reproductive division of labor (in ants and other eusocial **Hymenoptera** between reproductive **queens** and nonreproductive **workers**), cooperative **brood** care, and at least two overlapping adult generations that contribute to the **colony** (e.g., a mother **queen** and her daughter **workers**).

Exocrine gland: A gland that secretes substances onto the surface of an epithelium, often to the outside of the body. Most **pheromones** are derived from exocrine glands.

Exoskeleton: The external skeleton that shields and supports the body of an insect.

Family: The term usually refers to a group of related individuals. In **taxonomy**, however, it denotes the major rank above **genus**, and ideally represents a **clade** of closely related **genera**. For example, the family **Formicidae** includes all the ants and no other organisms.

Femur: Often the largest, but not necessarily longest, segment of the insect leg. It is particularly well developed in jumping insects like locusts and katydids, where it holds much of the musculature necessary for leaping (see Figure P.6).

Fission: See **colony fission**.

Fitness: The relative contribution of an organism (or **allele**, or **phenotype**) to the gene pool of the next generation; see also **inclusive fitness**. Note that this is only a verbal definition for the purpose of this book, rather than the precise technical definition in the framework of population genetics.

Formicidae: The ants, an insect **family** with over 14,000 living **species**. From the Latin word *formica* for ant.

Gaster: The last portion of an ant's body, posterior to the **petiole** or **postpetiole** (see Figure P.6).

Genotype: The combination of **alleles** an individual carries across its genome.

Genus (plural **genera):** In **taxonomy**, the rank above **species**. A genus contains closely related and similar **species** and should ideally represent a **clade**. The genus name is represented as the first part of a binomial **species** name, such as the genus *Eciton* in the **species** name *Eciton burchellii*.

Grooming: Cleaning the body surface by, for instance, licking or stroking. Animals can either groom themselves or social partners.

Group raid: A type of raid that likely was the evolutionary precursor of **army ant mass raids**, and that is employed by extant relatives of **army ants**. Group raids typically contain less than a thousand **workers**, and the raiding party does not remain connected to the **nest** via a continuous column of **workers**. In many group raiding **species**, colonies send out individual **scouts**, which then **recruit** raiding parties upon discovering prey. Raiding parties thus leave the **nest** and travel to the destination indicated by the **scout**.

Guest: An organism that lives in **symbiosis** with a **host**, either as a parasite (**parasitism**), commensal (**commensalism**), or mutualist (**mutualism**).

Haplodiploidy: A genetic system in which males are **haploid** and females are **diploid**, as is the case in all **Hymenoptera**.

Haploid: Having only a single complement of chromosomes, which is usually inherited from the mother.

Hemimetabolism: A mode of development that includes egg, nymph, and adult stages, but no **pupa** stage. The nymph often resembles the adult in general appearance. Hemimetabolism is also called incomplete **metamorphosis**. Termites, cockroaches, and locusts, for example, are all hemimetabolous insects.

Hemolymph: The fluid circulating inside the body of an arthropod, analogous to blood in vertebrates.

Holometabolism: A mode of development that includes egg, **larva** (or grub / maggot / caterpillar), **pupa** (or chrysalis), and adult stages. The **larva** does not resemble the adult. Holometabolism is also called complete **metamorphosis**. Ants, beetles, butterflies, and flies, for example, are all holometabolous insects.

Host: An organism that harbors a **guest**. In the context of social insects, the term can either refer to the individual insect, for example an ant infected by an **endoparasite**, or the **colony**, for example an **army ant colony** infected by a **social parasite**.

Hymenoptera: An **order** of insects comprised of the suborders Symphyta (the sawflies) and **Apocrita**.

Hymenopteran: Pertaining to the insect **order Hymenoptera**.

Hypogaeic: Active below ground / in the soil; most **army ants** conduct hypogaeic **mass raids** and **emigrations**, and construct their **bivouacs** hypogaeically. Only a few **species** regularly perform these activities **epigaeically**.

Imago: An insect at the developmental stage where it reaches sexual maturity; also called the adult. In **holometabolous** insects, the imago follows the **pupa**.

Inclusive fitness: The sum of an organism's direct and indirect **fitness**, via its own offspring and the offspring of related individuals, respectively.

Independent colony founding: Queens found new **colonies** in the absence of **nestmate workers**.

Inquiline: An organism that specifically lives inside ant **colonies**. In the **myrmeco-phile** literature, the term refers to organisms other than ants that live inside ant **nests** for at least part of their lifecycle. In the ant literature, the term denotes socially parasitic ants that spend their entire lifecycle exploiting **colonies** of **host** ant **species**.

Instar: A developmental stage of an **arthropod**, separated from other developmental stages by **molts**. For example, an ant **larva** will transition between different instars by **molting** several times before reaching the **pupal** stage.

Isometric scaling: A special case in **allometry**, where proportional relationships between two body parts remain constant as the overall size of the organism changes.

Kleptoparasitism: A type of **parasitism** in which one animal steals and eats the food collected by another animal.

Larva (plural **larvae):** An immature developmental stage of **holometabolous** insects that looks strikingly different from the **imago**, or adult. Depending on the type of insect, the larva is also called a grub, maggot, or caterpillar. After the larva hatches from the egg, it can transition through several larval **instars** that are separated by **molts**, before it enters the **pupal** stage and undergoes **metamorphosis**.

Larviparous: The **larva** hatches inside the female, which then gives birth to live **larvae** rather than laying eggs.

Limuloid: Drop-shaped. Literally, shaped like a horseshoe crab (**genus** *Limulus*).

Major: See **soldier**.

Mandible: A paired insect mouthpart often used for grasping and processing food, or for fighting and defense (see Figure P.6).

Mass raid: The type of raid employed by **army ants**. The raid is initiated spontaneous-ly at the **nest**, and leaves the **nest** without a predetermined destination. Prey is encountered as the raid progresses, and the raid front typically remains connected to the **nest** via a continuous column of **workers**. Mass raids usually contain more than 10,000 **workers**. This form of foraging likely evolved from **group raids**.

Meconium: In ants, the metabolic waste product expelled by the **larva** before becom-ing a **pupa**—that is, entering **metamorphosis**.

Melanin: A type of naturally occurring pigment. In many insects, the amount of melanin in the **cuticle** increases as a function of adult age, resulting in progressive darkening, or melanization.

Metamorphosis: A period in the development of an animal during which its body undergoes major remodeling. Some animals are **holometabolous** (they undergo complete metamorphosis), others are **hemimetabolous** (they undergo incomplete metamorphosis), and still others are ametabolous (they do not undergo metamor-phosis).

Mimicry: An adaptive, superficial resemblance of two organisms that are not closely related **phylogenetically**. Mimicry can be beneficial for both the mimic and the model (a case of **mutualism**), or benefit only the mimic. Different types of mimicry are Batesian mimicry, Müllerian mimicry, and Wasmannian mimicry.

Miocene: A geological epoch extending from ca. 23 to 5 million years ago.

Monandry: In **social insects**, a mating system in which a female mates with only one male, or fertilizes eggs with sperm from a single male.

Monogyny: In **social insects**, a type of **colony** organization where only one reproductively active **queen** is present at any one time.

Monomorphic: Being morphologically uniform.

Molt: In **arthropods**, the shedding of the outgrown skin, or **exoskeleton**. The term is also used as a noun to refer to the empty skin itself.

Mutualism: A type of **symbiosis** between two **species** from which both partners have a net **fitness** gain.

Mutualist: See **mutualism**.

Myrmecoid: Ant-like, usually in morphology or behavior.

Myrmecology: The scientific study of ants. The term is derived from the Greek words *myrmex* for ant, and *logos* for study.

Myrmecophagous: Feeding on ants.

Myrmecophile: An organism that lives in specific association with ants for at least part of its lifecycle. The term is derived from the Greek words *myrmex* for ant, and *philos* for loving.

Neotropics: The biogeographic realm constituting the New World / American tropics.

Nest: In **social insects**, the place where a **colony** lives. The nests of many **social insects** are elaborate, permanent constructions, but **army ants** build temporary **bivouac** nests instead.

Nestmate: An organism that inhabits the same **nest** as a focal individual. In **social insects**, the term usually refers to individuals from the same **colony**.

Nomadic phase: The part of the **phasic colony** cycle in which **larvae** are present, no eggs are being laid, and **colonies** show elevated foraging activity and frequent **emigrations** (see Figures 4.26 and 4.27).

Nomadism: Colonies conduct frequent **emigrations** to new nest sites.

Nonphasic: Species that are not **phasic**.

Order: In **taxonomy**, the major rank above **family**.

Ovariole: One of the filaments that form the insect ovary.

Parasite: See **parasitism**.

Parasitic: See **parasitism**.

Parasitism: A type of **symbiosis** between two **species** in which one organism, the parasite, lives on (**ectoparasitism**) or inside (**endoparasitism**) a **host**, which it exploits. While the parasite benefits from this relationship, the **host** bears a net **fitness** cost. See also **social parasitism.**

Parasitoid: A type of parasite (**parasitism**) that ultimately kills the **host**.

Petiole: The constricted segment of an ant's body forming the "wasp waist," joining the **alitrunk** to the **gaster** (see Figure P.6). In some ants, the petiole is immediately followed by a second constricted segment, the **postpetiole**, which then connects to the **gaster**.

Phasic: Colonies undergo stereotypical cycles, alternating between **statary phases** (also called **reproductive phases**) and **nomadic phases** (also called **brood care phases**). **Brood** develops in discrete cohorts, and **colony** behavior changes predictably throughout the cycle (see Figures 4.26 and 4.27).

Phenotype: The observable features of an organism. These include, among others, behavioral, developmental, morphological, and physiological traits. Simply put, the phenotype arises from interactions between an organism's **genotype** and environmental factors. Phenotypes can also be measured at the level of the **colony**.

Pheromone: A secreted or excreted chemical compound that elicits a behavioral or physiological response in members of the same **species**.

Phoresis: An often rather transient type of **commensalism** in which an organism attaches to a **host** for transport.

Phylogenetically: See **phylogeny**.

Phylogeny: The evolutionary relationships within a group of different entities, such as biological **species**, **genera**, or **subfamilies**, often depicted as a branching "tree" diagram. The entities available for study are shown as external "tips" of the tree, while their evolutionary relationships are inferred from the collected data, such as morphological characters or DNA sequences. Internal branching points, or "nodes," represent inferred common ancestors of the entities at the tips.

Physogastry: A temporary swelling of the **gaster** during times of increased egg-laying activity, to an extent that the membranous regions between the **sclerites** become visible.

Polyandry: In **social insects**, a mating system in which a female mates with more than one male, and fertilizes eggs with sperm from several males.

Polyethism: See **age polyethism**.

Polygyny: In social insects, a type of **colony** organization where several **queens** reproduce simultaneously.

Polymorphic: Being morphologically diverse.

Ponerinae: An ant **subfamily** that contains several species with army ant–like behavior as well as a few **army ants**.

Ponerine: Pertaining to the ant **subfamily Ponerinae**.

Postpetiole: A constricted segment of the body of some ants between the **petiole** and the **gaster** (see Figure P.6).

Preadaptation: A trait that has initially evolved for a different purpose but can be readily coopted for a different or expanded function under a new selective regime.

Pupa (plural **pupae):** The developmental stage of a **holometabolous** insect that undergoes **metamorphosis**. In butterflies this stage is called the chrysalis.

Pupation: The transition from the **larva** to the **pupa** during development.

Queen: The reproductive **caste** in an ant **colony**. In ant **species** with morphologically distinct female **castes**, queens are the largest individuals on average.

Queenright: Referring to a **colony** with a functional **queen**.

Recruitment: A behavior that directs members of a **colony** to a particular location. Recruitment can involve **pheromones** or other cues.

Reproductive phase: See **statary phase**.

Scape: See **antennal scape**.

Sclerite: A hardened region of the **arthropod exoskeleton**.

Scout: An individual that explores outside the nest to obtain information about features of the external environment, such as the location of food or alternative nesting sites.

Seta (plural **setae):** Any type of hairlike structure.

Sexual brood: A batch of **haploid** larvae that develop into males, and **diploid** larvae that develop into **queens**.

Sexual larva: See **sexual brood**.

Social insect: In this book, the term is used interchangeably with "**eusocial** insect."

Social parasite: See **social parasitism**.

Social parasitism: A relationship in which an organism lives inside a **eusocial colony** and exploits the social fabric of its **host**.

Society: See **colony** (the two terms are used interchangeably in this book).

Soldier: A **caste** of **worker** specialized on **colony** defense. The largest **worker caste** in **army ants**, also referred to as **major**.

Species: Members of a biological species can interbreed and produce fertile offspring, while members of different species cannot. Note, however, that this is a simplified definition for the purpose of this book, and that there is a large body of literature on different ways to define biological species. The species is also the basic unit in **taxonomy**.

Spermatheca: A sperm-storage organ in females. Especially in **eusocial Hymenoptera**, females mate only early in life and then store the acquired sperm for extended periods, often for years.

Statary phase: The part of the **phasic colony** cycle in which **pupae** are present, eggs are laid, and **colonies** show decreased foraging activity and do not conduct **emigrations** (see Figures 4.26 and 4.27).

Subfamily: In **taxonomy**, an intermediate rank between **genus** and **family**.

Submajor: A porter **caste** in *Eciton* **army ants** specialized on carrying large and bulky prey and **brood** items. The second largest **worker caste** after the **majors** / **soldiers**.

Subsocial: See **subsociality**.

Subsociality: A form of social living in which adults provide extended care for their young.

Superorganism: An insect **society** in which natural selection acts predominantly at the level of the **colony**, rather than the individual insect. From an evolutionary point of view, the **colony** behaves as an individual. By this definition, permanently fixed and morphologically differentiated **queen** and **worker castes**, which are now analogous to the germ-line and soma in metazoans, are required for the **colonies** of a given ant **species** to be classified as superorganismal.

Swarm raid: A type of **mass raid** in which **army ants** fan out from the **nest** in a carpet-like swarm, and raiding columns form in the wake of the swarm. As opposed to **column raids**, swarm raids only occur in **army ants** with extremely large **colony** sizes (see Figure 3.8).

Symbiosis: An intimate, persistent, and specific ecological interaction between the members of two different **species**. The three main types of symbiosis are **parasitism**, **commensalism**, and **mutualism**.

Tarsal claws: Paired claws at the distal end of the **tarsus** (see Figure P.6).

Tarsus: The part of the insect leg distal to the **tibia** (see Figure P.6); the "foot" of an insect. In most insects, the tarsus is divided into subsegments, or tarsomeres.

Taxonomy: The scientific discipline of identifying, classifying, describing, and naming organisms.

Tibia: The segment of the insect leg distal to the **femur** (see Figure P.6).

Trichome: A tuft of hairs associated with glandular openings on the body surface.

Trophallaxis: The transfer of liquid, predigested food between **colony** members, including **inquiline social parasites**.

Trophobiont: An organism that provides food to a partner in a **mutualism**, often in exchange for protection or other services. A **mutualism** involving a trophobiont is called a trophobiosis.

Worker: In **Hymenoptera**, a nonreproductive female that performs tasks like **brood** care, foraging, **nest** construction, and defense. In ant **species** with morphologically distinct female **castes**, the workers are smaller than the **queens**, on average.

references

Agrain FA, Buffington ML, Chaboo CS, Chamorro ML, Schöller M. 2015. Leaf beetles are ant-nest beetles: the curious life of the juvenile stages of case-bearers (Coleoptera, Chrysomelidae, Cryptocephalinae). *ZooKeys* 547: 133–164.

Akino T. 2008. Chemical strategies to deal with ants: a review of mimicry, camouflage, propaganda, and phytomimesis by ants (Hymenoptera: Formicidae) and other arthropods. *Myrmecological News* 11: 173–181.

Akre RD. 1968. The behavior of *Euxenister* and *Pulvinister,* histerid beetles associated with army ants. *Pan-Pacific Entomologist* 44: 87–101.

Akre RD, Rettenmeyer CW. 1966. Behavior of Staphylinidae associated with army ants (Formicidae: Ecitonini). *Journal of the Kansas Entomological Society* 39: 745–782.

Akre RD, Rettenmeyer CW. 1968. Trail-following by guests of army ants (Hymenoptera: Formicidae: Ecitonini). *Journal of the Kansas Entomological Society* 41: 165–174.

Akre RD, Torgerson RL. 1968. The behavior of *Diploeciton nevermanni,* a staphylinid beetle associated with army ants. *Psyche* 75: 211–215.

Akre RD, Torgerson RL. 1969. Behavior of *Vatesus* beetles associated with army ants (Coleoptera: Staphylinidae). *Pan-Pacific Entomologist* 45: 269–281.

Anonymous. 1701. Diverses observations de physique générale. Histoire de l'Académie Royale des Sciences de Paris, Année MDCCI: 16. In *Abrégé de l'Histoire et des Mémoires de l'Académie Royale des Sciences, Tome Premier,* edited by G de Montbeliard, 634. Paris: self-published.

Arimoto K, Yamane S. 2018. Taxonomy of the *Leptogenys chalybaea* species group (Hymenoptera, Formicidae, Ponerinae) from Southeast Asia. *Asian Myrmecology* 10: e010008.

Attygalle AB, Vorstrowsky O, Bestmann HJ, Steghaus-Kovac S, Maschwitz U. 1988. (3*R*, 4*S*)-4-methyl-3-heptanol, the trail pheromone of the army ant *Leptogenys diminuta*. *Naturwissenschaften* 75: 315–317.

Auerbach, J. 2014. Does New York City really have as many rats as people? *Significance* 11: 22–27.

Bagneres AG, Billen J, Morgan ED. 1991. Volatile secretion of Dufour gland of workers of an army ant, *Dorylus (Anomma) molestus. Journal of Chemical Ecology* 17: 1633–1639.

Baldridge RS, Rettenmeyer CW, Watkins JF II. 1980. Seasonal, nocturnal and diurnal flight periodicities of Nearctic army ant males (Hymenoptera: Formicidae). *Journal of the Kansas Entomological Society* 53: 189–204.

Barden P. 2017. Fossil ants (Hymenoptera: Formicidae): ancient diversity and the rise of modern lineages. *Myrmecological News* 24: 1–30.

Barth MB, Moritz RFA, Pirk CWW, Kraus FB. 2013. Male-biased dispersal promotes large scale gene flow in a subterranean army ant, *Dorylus (Typhlopone) fulvus. Population Ecology* 55: 523–533.

Barth MB, Moritz RFA, Kraus FB. 2014. The evolution of extreme polyandry in social insects: insights from army ants. *PLoS One* 9: e105621.

Bartholomew GA, Lighton JRB, Feener DH Jr. 1988. Energetics of trail running, load carriage, and emigration in the column-raiding army ant *Eciton hamatum. Physiological Zoology* 61: 57–68.

Bates HW. 1862. Contributions to an insect fauna of the Amazon Valley. *Transactions of the Linnean Society of London* 23: 495–566.

Bates HW. 1863. *The Naturalist on the River Amazons.* London: J. M. Dent & Sons.

Baudier KM, O'Donnell S. 2016. Structure and thermal biology of subterranean army ant bivouacs in a tropical montane forest. *Insectes Sociaux* 63: 467–476.

Baudier KM, O'Donnell S. 2018. Complex body size differences in thermal tolerance among army ant workers (*Eciton burchellii parvispinum*). *Journal of Thermal Biology* 78: 277–280.

Baudier KM, Mudd AE, Erickson SC, O'Donnell S. 2015. Microhabitat and body size effects on heat tolerance: implications for responses to climate change (army ants: Formicidae, Ecitoninae). *Journal of Animal Ecology* 84: 1322–1330.

Baudier KM, D'Amelio CL, Malhotra R, O'Connor MP, O'Donnell S. 2018. Extreme insolation: climatic variation shapes the evolution of thermal tolerance at multiple scales. *American Naturalist* 192: 347–359.

Baudier KM, D'Amelio CL, Sulger E, O'Connor MP, O'Donnell S. 2019. Plastic collective endothermy in a complex animal society (army ant bivouacs: *Eciton burchellii parvispinum*). *Ecography* 42: 730–739.

Bayliss J, Fielding A. 2002. Termitophagous foraging by *Pachycondyla analis* (Formicidae, Ponerinae) in a Tanzanian coastal dry forest. *Sociobiology* 39: 103–122.

Beck J, Kunz BK. 2007. Cooperative self-defence: Matabele ants (*Pachycondyla analis*) against African driver ants (*Dorylus* sp.; Hymenoptera: Formicidae). *Myrmecological News* 10: 27–28.

Beebe W. 1917. With army ants "somewhere" in the jungle. *Atlantic Monthly,* April 1917: 514–522.

Beebe W. 1919. The home town of the army ants. *Atlantic Monthly,* October 1919: 454–464.

Beebe W. 1921. *Edge of the Jungle.* New York: Henry Holt.

Belt T. 1874. *The Naturalist in Nicaragua.* London: John Murray.

Berghoff SM, Franks NR. 2007. First record of the army ant *Cheliomyrmex morosus* in Panama and its high associate diversity. *Biotropica* 39: 771–773.

Berghoff SM, Weissflog A, Linsenmair KE, Hashim R, Maschwitz U. 2002a. Foraging of a hypogaeic army ant: a long neglected majority. *Insectes Sociaux* 49: 133–141.

Berghoff SM, Weissflog A, Linsenmair KE, Mohamed M, Maschwitz U. 2002b. Nesting habits and colony composition of the hypogaeic army ant *Dorylus* (*Dichthadia*) *laevigatus* Fr. Smith. *Insectes Sociaux* 49: 380–387.

Berghoff SM, Gadau J, Winter T, Linsenmair KE, Maschwitz U. 2003a. Sociobiology of hypogaeic army ants: characterization of two sympatric *Dorylus* species on Borneo and their colony conflicts. *Insectes Sociaux* 50: 139–147.

Berghoff SM, Maschwitz U, Linsenmair KE. 2003b. Influence of the hypogaeic army ant *Dorylus* (*Dichthadia*) *laevigatus* on tropical arthropod communities. *Oecologia* 135: 149–157.

Berghoff SM, Kronauer DJC, Edwards KJ, Franks NR. 2008. Dispersal and population structure of a New World predator, the army ant *Eciton burchellii*. *Journal of Evolutionary Biology* 21: 1125–1132.

Berghoff SM, Wurst E, Ebermann E, Sendova-Franks AB, Rettenmeyer CW, Franks N. 2009. Symbionts of societies that fission: mites as guests or parasites of army ants. *Ecological Entomology* 34: 684–695.

Beye M, Hasselmann M, Fondrk MK, Page RE Jr, Omholt SW. 2003. The gene *csd* is the primary signal for sexual development in the honeybee and encodes an SR-type protein. *Cell* 114: 419–429.

Billen J. 1992. Origin of the trail pheromone in Ecitoninae: a behavioural and morphological examination. In *Biology and Evolution of Social Insects,* edited by J Billen, 203–209. Leuven, the Netherlands: Leuven University Press.

Billen J, Gobin B. 1996. Trail following in army ants (Hymenoptera, Formicidae). *Netherlands Journal of Zoology* 46: 272–280.

Blum MS, Portocarrero CA. 1964. Chemical releaser of social behavior. IV. The hindgut as the source of the odor trail pheromone in the Neotropical army ant genus *Eciton*. *Annals of the Entomological Society of America* 57: 793–794.

Blüthgen N, Fiedler K. 2002. Interactions between weaver ants *Oecophylla smaragdina*, homopterans, trees and lianas in an Australian rain forest canopy. *Journal of Animal Ecology* 71: 793–801.

Bodot P. 1961. La destruction des termitières de *Bellicositermes natalensis* Hav., par une Fourmi, *Dorylus* (*Typhlopone*) *dentifrons* Wasmann. *Comptes Rendus Hebdomadaires des Séances de l'Académie des Sciences* 253: 3053–3054.

Bollag B. 2001. Word by word, researchers build an ode to Swahili poetry. *Chronicle of Higher Education,* May 25, 2001: A56.

Bolton B, Fisher BL. 2012. Taxonomy of the cerapachyine ant genera *Simopone* Forel, *Vicinopone* gen. n. and *Tanipone* gen. n. (Hymenoptera: Formicidae). *Zootaxa* 3283: 1–101.

Boomsma JJ, Gawne R. 2018. Superorganismality and caste differentiation as points of no return: how the major evolutionary transitions were lost in translation. *Biological Reviews* 93: 28–54.

Boomsma JJ, Kronauer DJC, Pedersen JS. 2009. The evolution of social insect mating systems. In *Organization of Insect Societies: From Genome to Sociocomplexity,* edited by J Gadau and J Fewell, 3–25. Cambridge, MA: Harvard University Press.

Borgmeier T. 1930. Zwei neue Gattungen ecitophiler Aleocharinen (Col., Staph.). *Zoologischer Anzeiger* 92: 165–178.

Borgmeier T. 1953. Vorarbeiten zu einer Revision der neotropischen Wanderameisen. *Studia Entomologica* 2: 1–51.

Borgmeier T. 1955. Die Wanderameisen der neotropischen Region (Hym. Formicidae). *Studia Entomologica* 3: 1–716.

Borgmeier T. 1958. Nachtraege zu meiner Monographie der neotropischen Wanderameisen (Hym. Formicidae). *Studia Entomologica* 1 (1–2): 197–208.

Borgmeier T. 1963. New or little known *Coniceromyia,* and some other Neotropical or Paleotropical Phoridae (Diptera). *Studia Entomologica* 6: 449–480.

Borowiec ML. 2009. New species related to *Cerapachys sexspinus* and discussion of the status of *Yunodorylus. Zootaxa* 2069: 43–58.

Borowiec ML. 2016. Generic revision of the ant subfamily Dorylinae (Hymenoptera, Formicidae). *ZooKeys* 608: 1–280.

Borowiec ML. 2019. Convergent evolution of the army ant syndrome and congruence in big-data phylogenetics. *Systematic Biology* 68: 642–656.

Borowiec ML, Longino JT. 2011. Three new species and reassessment of the rare Neotropical ant genus *Leptanilloides* (Hymenoptera, Formicidae, Leptanilloidinae). *ZooKeys* 133: 19–48.

Borowiec ML, Rabeling C, Brady SG, Fisher BL, Schultz TR, Ward PS. 2019. Compositional heterogeneity and outgroup choice influence the internal phylogeny of the ants. *Molecular Phylogenetics and Evolution* 134: 111–121.

Boswell GP, Britton NF, Franks NR. 1998. Habitat fragmentation, percolation theory and the conservation of a keystone species. *Proceedings of the Royal Society of London, Series B: Biological Sciences* 265: 1921–1925.

Boswell GP, Franks NR, Britton NF. 2001. Arms races and the evolution of big fierce societies. *Proceedings of the Royal Society of London, Series B: Biological Sciences* 268: 1723–1730.

Bourke AFG. 1999. Colony size, social complexity and reproductive conflict in social insects. *Journal of Evolutionary Biology* 12: 245–257.

Bourke AFG, Franks NR. 1995. *Social Evolution in Ants.* Princeton, NJ: Princeton University Press.

Brady SG. 2003. Evolution of the army ant syndrome: the origin and long-term evolutionary stasis of a complex of behavioral and reproductive adaptations. *Proceedings of the National Academy of Sciences of the United States of America* 100: 6575–6579.

Brady SG, Ward PS. 2005. Morphological phylogeny of army ants and other dorylomorphs (Hymenoptera: Formicidae). *Systematic Entomology* 30: 593–618.

Brady SG, Schultz TR, Fisher BL, Ward PS. 2006. Evaluating alternative hypotheses for the early evolution and diversification of ants. *Proceedings of the National Academy of Sciences of the United States of America* 103: 18172–18177.

Brady SG, Fisher BL, Schultz TR, Ward PS. 2014. The rise of army ants and their relatives: diversification of specialized predatory doryline ants. *BMC Evolutionary Biology* 14: 93.

Braendle C, Hockley N, Brevig T, Shingleton AW, Keller L. 2003. Size-correlated division of labour and spatial distribution of workers in the driver ant, *Dorylus molestus. Naturwissenschaften* 90: 277–281.

Brandão CRF, Diniz JL, Agosti D, Delabie JH. 1999. Revision of the Neotropical ant subfamily Leptanilloidinae. *Systematic Entomology* 24: 17–36.

Brandão CRF, Machado Diniz JL, dos Santos Machado Feitosa R. 2010. The venom apparatus and other morphological characters of the ant *Martialis heureka* (Hymenoptera, Formicidae, Martialinae). *Papéis Avulsos de Zoologia (São Paulo)* 50: 413–423.

Branstetter MG, Danforth BN, Pitts JP, Faircloth BC, Ward PS, Buffington ML, Gates MW, Kula RR, Brady SG. 2017a. Phylogenomic insights into the evolution of stinging wasps and the origins of ants and bees. *Current Biology* 27: 1019–1025.

Branstetter MG, Ješovnik A, Sosa-Calvo J, Lloyd MW, Faircloth BC, Brady SG, Schultz TR. 2017b. Dry habitats were crucibles of domestication in the evolution of agriculture in ants. *Proceedings of the Royal Society of London, Series B: Biological Sciences* 284: 20170095.

Briese DT. 1984. Interactions between a myrmecophagous ant and a prey species. *Journal of the Australian Entomological Society* 23: 167–168.

Britton NF, Partridge LW, Franks NR. 1996. A mathematical model for the population dynamics of army ants. *Bulletin of Mathematical Biology* 58: 471–492.

Britton NF, Boswell GP, Franks NR. 2001. Dispersal and conservation in heterogeneous landscapes. In *Insect Movement: Mechanisms and Consequences,* edited by IP Woiwod, DR Reynolds, and CD Thomas, 299–320. Wallingford, United Kingdom: CAB International.

Broadley DG, Wallach V. 2009. A review of the eastern and southern African blind-snakes (Serpentes: Typhlopidae), excluding *Letheobia* Cope, with the description of two new genera and a new species. *Zootaxa* 2255 (1): 1–100.

Brooks M. 2006. *World War Z: An Oral History of the Zombie War.* New York: Crown.

Brown WL Jr. 1960. The release of alarm and attack behavior in some New World army ants. *Psyche* 66: 25–27.

Brown WL Jr. 1975. Contributions toward a reclassification of the Formicidae. V. Ponerinae, tribes Platythyreini, Cerapachyini, Cylindromyrmecini, Acanthostichini, and Aenictogitini. *Search: Agriculture* (Cornell University) 5 (1): 1–115.

Brown BV. 2017. Fossil evidence of social insect commensalism in the Phoridae (Insecta: Diptera). *Journal of Systematic Palaeontology* 15: 275–285.

Brown BV, Feener DH Jr. 1998. Parasitic phorid flies (Diptera: Phoridae) associated with army ants (Hymenoptera: Formicidae: Ecitoninae, Dorylinae) and their conservation biology. *Biotropica* 30: 482–487.

Brown CA, Watkins JF II, Eldridge DW. 1979. Repression of bacteria and fungi by the army ant secretion: skatole. *Journal of the Kansas Entomological Society* 52: 119–122.

Brown BV, Hayes C, Hash JM, Smith PT. 2018. Molecular phylogeny of the ant-decapitating flies, genus *Apocephalus* Coquillett (Diptera: Phoridae). *Insect Systematics and Diversity* 2 (4): 2.

Bruch C. 1923. Estudios mirmecológicos—con la descripción de nuevas especies de dípteros (Phoridae) por los Rr. Pp. H. Schmitz y Th. Borgmeier y de una araña (Gonyleptidae) por el doctor Mello-Leitao. *Revista del Museo de La Plata* 27: 172–220.

Brückner A, Klompen H, Bruce AI, Hashim R, von Beeren C. 2017. Infection of army ant pupae by two new parasitoid mites (Mesostigmata: Uropodina). *PeerJ* 5: e3870.

Brückner A, Hoenle PO, von Beeren C. 2018. Comparative chemical analysis of army ant mandibular gland volatiles (Formicidae: Dorylinae). *PeerJ* 6: e5319.

Brumfield RT, Tello JG, Cheviron ZA, Carling MD, Crochet N, Rosenberg KV. 2007. Phylogenetic conservatism and antiquity of a tropical specialization: army-ant-following in the typical antbirds (Thamnophilidae). *Molecular Phylogenetics and Evolution* 45: 1–13.

Bulmer MG. 1983. Sex ratio theory in social insects with swarming. *Journal of Theoretical Biology* 100: 329–339.

Bulova S, Purce K, Khodak P, Sulger E, O'Donnell S. 2016. Into the black and back: the ecology of brain investment in Neotropical army ants (Formicidae: Dorylinae). *Science of Nature* 103: 31.

Burton JL, Franks NR. 1985. The foraging ecology of the army ant *Eciton rapax:* an ergonomic enigma? *Ecological Entomology* 10: 131–141.

Buschinger A, Peeters C, Crozier RH. 1989. Life-pattern studies on an Australian *Sphinctomyrmex* (Formicidae: Ponerinae; Cerapachyini): functional polygyny, brood periodicity and raiding behavior. *Psyche* 96: 287–300.

Butler IA, Peters MK, Kronauer DJC. 2018. Low levels of hybridization in two species of African driver ants. *Journal of Evolutionary Biology* 31: 556–571.

Califano D, Chaves-Campos J. 2011. Effect of trail pheromones and weather on the moving behaviour of the army ant *Eciton burchellii. Insectes Sociaux* 58: 309–315.

Campbell KU, Klompen H, Crist TO. 2013. The diversity and host specificity of mites associated with ants: the roles of ecological and life history traits of ant hosts. *Insectes Sociaux* 60: 31–41.

Cansdale GS. 1961. *West African Nature Handbooks: West African Snakes.* London: Longman.

Carroll L. 1871. *Through the Looking-Glass.* London: Macmillan and Co.

Caterino MS, Tishechkin AK. 2006. DNA identification and morphological description of the first confirmed larvae of Hetaeriinae (Coleoptera: Histeridae). *Systematic Entomology* 31: 405–418.

Cerdá X, Dejean A. 2011. Predation by ants on arthropods and other animals. In *Predation in the Hymenoptera: An Evolutionary Perspective,* edited by C Polidori, 39–78. Kerala, India: Transworld Research Network.

Chadab R. 1979. Early warning cues for social wasps attacked by army ants. *Psyche* 86: 115–124.

Chadab R, Rettenmeyer CW. 1975. Mass recruitment by army ants. *Science* 188: 1124–1125.

Chadab-Crepet R, Rettenmeyer CW. 1982. Comparative behaviour of social wasps when attacked by army ants or other predators and parasites. In *The Biology of Social Insects,* edited by MD Breed, CO Michener, and HE Evans, 270–274. Boulder, CO: Westview Press.

Chandra V, Fetter-Pruneda I, Oxley PR, Ritger AL, McKenzie SK, Libbrecht R, Kronauer DJC. 2018. Social regulation of insulin signaling and the evolution of eusociality in ants. *Science* 361: 398–402.

Chandra V, Gal A, Kronauer DJC. n.d. Colony size expansions can explain the evolution of army ant mass raiding. Unpublished manuscript.

Chapman JW. 1964. Studies on the ecology of the army ants of the Philippines genus *Aenictus* Shuckard (Hymenoptera: Formicidae). *Philippine Journal of Science* 93: 551–595.

Chaves-Campos J. 2003. Localization of army-ant swarms by ant-following birds on the Caribbean slope of Costa Rica: following the vocalization of antbirds to find the swarms. *Ornitología Neotropical* 14: 289–294.

Chaves-Campos J. 2011. Ant colony tracking in the obligate army ant-following antbird *Phaenostictus mcleannani. Journal of Ornithology* 152: 497–504.

Chaves-Campos J, DeWoody JA. 2008. The spatial distribution of avian relatives: do obligate army-ant-following birds roost and feed near family members? *Molecular Ecology* 17: 2963–2974.

Chesser RT. 1995. Comparative diets of obligate ant-following birds at a site in northern Bolivia. *Biotropica* 27: 382–390.

Clark J. 1923. Australian Formicidae. *Journal of the Royal Society of Western Australia* 9: 72–89.

Clark J. 1924. Australian Formicidae. *Journal of the Royal Society of Western Australia* 10: 75–89.

Clark J. 1941. Australian Formicidae. Notes and new species. *Memoirs of the National Museum, Victoria* 12: 71–94.

Coates-Estrada R, Estrada A. 1989. Avian attendance and foraging at army-ant swarms in the tropical rain forest of Los Tuxtlas, Veracruz, Mexico. *Journal of Tropical Ecology* 5: 281–292.

Cody ML. 2000. Antbird guilds in the lowland Caribbean rainforest of southeast Nicaragua. *The Condor* 102: 784–794.

Cohic F. 1948. Observations morphologiques et écologiques sur *Dorylus* (*Anomma*) *nigricans* Illiger (Hymenoptera Dorylidae). *Revue Française d'Entomologie* 14: 229–275.

Corbara B, Dejean A. 2000. Adaptive behavioral flexibility of the ant *Pachycondyla analis* (= *Megaponera foetens*) (Formicidae: Ponerinae) during prey capture. *Sociobiology* 36: 465–483.

Corbara B, Servigne P, Dejean A, Carpenter JM, Orivel J. 2018. A mimetic nesting association between a timid social wasp and an aggressive arboreal ant. *Comptes Rendus Biologies* 341: 182–188.

Coty D, Aria C, Garrouste R, Wils P, Legendre F, Nel A. 2014. The first ant-termite syninclusion in amber with CT-scan analysis of taphonomy. *PLoS One* 9: e104410.

Couzin ID, Franks NR. 2003. Self-organized lane formation and optimized traffic flow in army ants. *Proceedings of the Royal Society of London, Series B: Biological Sciences* 270: 139–146.

Cronin AL, Molet M, Doums C, Monnin T, Peeters C. 2013. Recurrent evolution of dependent colony foundation across eusocial insects. *Annual Review of Entomology* 58: 37–55.

Crozier RH, Pamilo P. 1996. *Evolution of Social Insect Colonies: Sex Allocation and Kin Selection.* Oxford: Oxford University Press.

Cushing PE. 2012. Spider-ant associations: an updated review of myrmecomorphy, myrmecophily, and myrmecophagy in spiders. *Psyche* 2012: 151989.

Darlington JPEC 1985 Attacks by doryline ants and termite nest defences (Hymenoptera: Formicidae; Isoptera: Termitidae). *Sociobiology* 11: 189–200.

Darwin C. 1839. *Narrative of the surveying voyages of His Majesty's ships* Adventure *and* Beagle *between the years 1826 and 1836, describing their examination of the southern shores of South America, and the* Beagle*'s circumnavigation of the globe. Journal and remarks. 1832–1836.* London: Henry Colburn.

Darwin C. 1859. *On the Origin of Species by Means of Natural Selection, or the Preservation of Favoured Races in the Struggle for Life.* London: John Murray.

De Andrade ML. 1998. Fossil and extant species of *Cylindromyrmex* (Hymenoptera: Formicidae). *Revue Suisse de Zoologie* 105: 581–664.

De Araújo Esteves F, Brandão CRF, Viegas K. 2008. Subterranean ants (Hymenoptera, Formicidae) as prey of fossorial reptiles (Reptilia, Squamata: Amphisbaenidae) in Central Brazil. *Papéis Avulsos de Zoologia (São Paulo)* 48: 329–334.

Dejean A, Corbara B. 2014. Reactions by ant workers to nestmates having had contact with sympatric ant species. *Comptes Rendus Biologies* 337: 642–645.

Dejean A, Evraerts C. 1997. Predatory behavior in the genus *Leptogenys:* a comparative study. *Journal of Insect Behavior* 10: 177–191.

Dejean A, Orivel J, Corbara B, Olmsted I, Lachaud JP. 2001. Nest site selection by two polistine wasps: the influence of *Acacia-Pseudomyrmex* associations against predation by army ants (Hymenoptera). *Sociobiology* 37: 135–146.

Dejean A, Corbara B, Roux O, Orivel J. 2013. The antipredatory behaviours of Neotropical ants towards army ant raids (Hymenoptera: Formicidae). *Myrmecological News* 19: 17–24.

De Melo Júnior TA, Zara FJ. 2007. Black-tufted-ear marmoset *Callithrix penicillata* (Primates: Callitrichidae) following the army ant *Labidus praedator* (Formicidae: Ecitoninae) in the Cerrado and the Atlantic forest, Brazil. *Neotropical Primates* 14: 32–33.

Deneubourg JL, Goss S, Franks NR, Pasteels JM. 1989. The blind leading the blind: modeling chemically mediated army ant raid patterns. *Journal of Insect Behavior* 2: 719–725.

Denny AJ, Franks NR, Powell S, Edwards KJ. 2004. Exceptionally high levels of multiple mating in an army ant. *Naturwissenschaften* 91: 396–399.

De Souza FSC, Calouro AM. 2018. Predation of army ants by Toppin's titi monkey, *Plecturocebus toppini* Thomas 1914 (Primates: Pitheciidae), in an urban forest fragment in eastern Acre, Brazil. *Primates* 59: 469–474.

Dill M, Williams DJ, Maschwitz U. 2002. Herdsmen ants and their mealybug partners. *Abhandlungen der Senckenbergischen Naturforschenden Gesellschaft Frankfurt am Main* 557: 1–373.

Disney RHL. 1994. *Scuttle Flies: The Phoridae.* London: Chapman & Hall.

Disney RHL. 1996. A new genus of scuttle fly (Diptera; Phoridae) whose legless, wingless, females mimic ant larvae (Hymenoptera; Formicidae). *Sociobiology* 27: 95–118.

Disney RHL, Berghoff SM. 2005. New species and new records of scuttle flies (Diptera: Phoridae) associated with army ants (Hymenoptera: Formicidae) in Trinidad and Venezuela. *Sociobiology* 45: 887–898.

Disney RHL, Berghoff SM. 2007. New species and new records of scuttle flies (Diptera: Phoridae) associated with army ants (Hymenoptera: Formicidae) in Panama. *Sociobiology* 49: 59–92.

Disney RHL, Berghoff SM. 2008. Further records, with a new species, of scuttle flies (Diptera: Phoridae) associated with army ants (Hymenoptera: Formicidae) in Panama. *Sociobiology* 51: 119–125.

Disney RHL, Kistner DH. 1990. The genus *Adelopteromyia*, which associates with Neotropical army ants (Diptera, Phoridae; Hymenoptera, Formicidae). *Sociobiology* 16: 259–263.

Disney RHL, Kistner DH. 1998. New species and new records of myrmecophilous Phoridae (Diptera). *Sociobiology* 31: 291–349.

Disney RHL, Kistner DH. 2003. New species and new host records of scuttle flies (Diptera: Phoridae) associated with army ants and termites (Hymenoptera: Formicidae; Isoptera: Termitidae). *Sociobiology* 42: 503–518.

Disney RHL, Rettenmeyer CW. 2007. New species and revisionary notes on scuttle flies (Diptera: Phoridae) associated with Neotropical army ants (Hymenoptera: Formicidae). *Sociobiology* 49: 1–58.

Disney RHL, Rettenmeyer CW. 2010. New species and new records of scuttle flies (Diptera: Phoridae) associated with Neotropical army ants (Hymenoptera: Formicidae). *Sociobiology* 55: 7–84.

Disney RHL, Weissflog A, Maschwitz U. 1998. A second species of legless scuttle fly (Diptera: Phoridae) associated with ants (Hymenoptera: Formicidae). *Journal of Zoology* 246: 269–274.

Disney RHL, Elizalde L, Folgarait PJ. 2008. New species and records of scuttle flies (Diptera: Phoridae) associated with leaf-cutter ants and army ants (Hymenoptera: Formicidae) in Argentina. *Sociobiology* 51: 95–117.

Disney RHL, Elizalde L, Folgarait PJ. 2009a. New species and new records of scuttle flies (Diptera: Phoridae) that parasitize leaf-cutter and army ants (Hymenoptera: Formicidae). *Sociobiology* 54: 601–631.

Disney RHL, Lizon à l'Allemand S, von Beeren C, Witte V. 2009b. A new genus and new species of scuttle flies (Diptera: Phoridae) from colonies of ants (Hymenoptera: Formicidae) in Malaysia. *Sociobiology* 53: 1–12.

Dobbs RC, Martin PR. 1998. Migrant bird participation at an army swarm in montane Jalisco, Mexico. *Wilson Bulletin* 10: 293–295.

Do Nascimento I, Delabie JHC, Fiúza Ferreira PS, Della Lucia TMC. 2004. Mating flight seasonality in the genus *Labidus* (Hymenoptera: Formicidae) at Minas Gerais, in the Brazilian Atlantic Forest Biome, and *Labidus nero,* junior synonym of *Labidus mars*. *Sociobiology* 44: 1–8.

Do Nascimento I, Delabie JHC, Della Lucia TMC. 2011. Phenology of mating flight in Ecitoninae (Hymenoptera: Formicidae) in a Brazilian Atlantic Forest location. *Annales de la Société Entomologique de France* 47: 112–118.

Donoso DA, Vieira JM, Wild AL. 2006. Three new species of *Leptanilloides* Mann from Andean Ecuador (Formicidae: Leptanilloidinae). *Zootaxa* 1201: 47–62.

Dornhaus A, Powell S. 2010. Foraging and defence strategies in ants. In *Ant Ecology,* edited by L Lach, CL Parr, and KL Abbott, 210–230. Oxford: Oxford University Press.

Dornhaus A, Powell S, Bengston S. 2012. Group size and its effects on collective organization. *Annual Reviews of Entomology* 57: 123–141.

Driver RJ, DeLeon S, O'Donnell S. 2018. Novel observation of a raptor, collared forest-falcon (*Micrastur semitorguatus*), depredating a fleeing snake at an army ant (*Eciton burchellii parvispinum*) raid front. *Wilson Journal of Ornithology* 130: 792–796.

Droual R. 1984. Anti-predator behaviour in the ant *Pheidole desertorum:* the importance of multiple nests. *Animal Behaviour* 32: 1054–1058.

Du Chaillu PB. 1861. *Explorations and Adventures in Equatorial Africa.* New York: Harper & Bros.

Duncan FD, Crewe RM. 1994. Group hunting in a ponerine ant, *Leptogenys nitida* Smith. *Oecologia* 97: 118–123.

Eguchi K, Bui VT, Oguri E, Maruyama M, Yamane S. 2014. A new data of worker polymorphism in the ant genus *Dorylus* (Hymenoptera: Formicidae: Dorylinae). *Journal of Asia-Pacific Entomology* 17: 31–36.

Eguchi K, Mizuno R, Ito F, Satria R, An DV, Viet BT, Luong PTH. 2016. First discovery of subdichthadiigyne in *Yunodorylus* Xu, 2000 (Formicidae: Dorylinae). *Revue Suisse de Zoologie* 123: 307–314.

Eickwort GC. 1990. Associations of mites with social insects. *Annual Review of Entomology* 35: 469–488.

Elmes GW. 1996. Biological diversity of ants and their role in ecosystem function. In *Biodiversity Research and Its Perspectives in East Asia,* edited by BH Lee, TH Kim, and BY Sun, 33–48. Proceedings of Inaugural Seminar of KIBIO. Korea: Chonbuk National University.

Elzinga RJ. 1978. Holdfast mechanisms in certain uropodine mites (Acarina: Uropodina). *Annals of the Entomological Society of America* 71: 896–900.

Elzinga RJ. 1993. Larvamimidae, a new family of mites (Acari: Dermanyssoidea) associated with army ants. *Acarologia* 34: 95–103.

Elzinga RJ. 1998. A new genus and five new species of mites (Acari: Ascidae) associated with army ants (Hymenoptera: Formicidae). *Sociobiology* 31: 351–361.

Elzinga RJ, Rettenmeyer CW. 1970. Five new species of *Planodiscus* (Acarina: Uropodina) found on doryline ants. *Acarologia* 12: 59–70.

Elzinga RJ, Rettenmeyer CW. 1974. Seven new species of *Circocylliba* (Acarina: Uropodina) found on army ants. *Acarologia* 16: 595–611.

Emery C. 1890. Studii sulle formiche della fauna neotropica. I–V. *Bollettino della Società Entomologica Italiana* 22: 38–80.

Emery C. 1896. Studi sulle formiche della fauna neotropica. XVII–XXV. *Bollettino della Società Entomologica Italiana* 28: 33–107.

Emlen DJ. 1996. Dung beetles unaffected by army ant swarm. Supplement, *Journal of the Kansas Entomological Society* 69: 405–406.

Fabricius JC. 1782. *Species insectorum exhibentes eorum differentias specificas, synonyma, auctorum loca natalia, metamorphosin adiectis observationibus, descriptionibus. Tome I.* Hamburg: C. E. Bohn.

Feener DH Jr. 1988. Effects of parasites on foraging and defense behavior of a termitophagous ant, *Pheidole titanis* Wheeler (Hymenoptera: Formicidae). *Behavioral Ecology and Sociobiology* 22: 421–427.

Feener DH Jr, Lighton JRB, Bartholomew GA. 1988. Curvilinear allometry, energetics and foraging ecology: a comparison of leaf-cutting ants and army ants. *Functional Ecology* 2: 509–520.

Fisher BL, Peeters C. 2019. *Ants of Madagascar: A Guide to the 62 Genera.* Antananarivo, Madagascar: Association Vahatra.

Ford AT, Goheen JR. 2015. Trophic cascades by large carnivores: a case for strong inference and mechanisms. *Trends in Ecology and Evolution* 30: 725–735.

Forel A. 1869. Observations sur les moeurs du *Solenopsis fugax. Mitteilungen der Schweizerischen Entomologischen Gesellschaft* 3: 105–128.

Forel A. 1886. Einige Ameisen aus Itajahy (Brasilien). *Mitteilungen der Schweizerischen Entomologischen Gesellschaft* 7: 210–217.

Forel A. 1937. *Out of My Life and Work.* London: George Allen & Unwin.

Fowler HG. 1977. Field response of *Acromyrmex crassispinus* (Forel) to aggression by *Atta sexdens* (Linn.) and predation by *Labidus praedator* (Fr. Smith) (Hymenoptera: Formicidae). *Aggressive Behavior* 3: 385–391.

Fowler HG. 1979. Notes on *Labidus praedator* (Fr. Smith) in Paraguay (Hymenoptera: Formicidae: Dorylinae: Ecitonini). *Journal of Natural History* 13: 3–10.

Frank ET. 2020. Matabele ant (*Megaponera analis*). In *Encyclopedia of Social Insects*, edited by CK Starr. Cham: Springer.

Frank ET, Linsenmair KE. 2017a. Flexible task allocation and raid organization in the termite-hunting ant *Megaponera analis. Insectes Sociaux* 64: 579–589.

Frank ET, Linsenmair KE. 2017b. Individual versus collective decision making: optimal foraging in the group hunting termite specialist *Megaponera analis. Animal Behaviour* 130: 27–35.

Frank ET, Schmitt T, Hovestadt T, Mitesser O, Stiegler J, Linsenmair KE. 2017. Saving the injured: rescue behavior in the termite-hunting ant *Megaponera analis. Science Advances* 3: e1602187.

Frank ET, Hönle PO, Linsenmair KE. 2018a. Time optimized path-choice in the termite hunting ant *Megaponera analis. Journal of Experimental Biology* 221: jeb174854.

Frank ET, Wehrhahn M, Linsenmair KE. 2018b. Wound treatment and selective help in a termite-hunting ant. *Proceedings of the Royal Society B: Biological Sciences* 285: 20172457.

Franks NR. 1982a. Ecology and population regulation in the army ant *Eciton burchelli.* In *The Ecology of a Tropical Forest: Seasonal Rhythms and Long-Term Changes,* edited by EG Leigh, AS Rand, and DM Windsor, 389–395. Washington, DC: Smithsonian Institution Press.

Franks NR. 1982b. A new method for censusing animal populations: the number of *Eciton burchelli* army ant colonies on Barro Colorado Island, Panama. *Oecologia* 52: 266–268.

Franks NR. 1985. Reproduction, foraging efficiency and worker polymorphism in army ants. In: *Fortschritte der Zoologie 31. Experimental Behavioral Ecology and Sociobiology: In Memoriam Karl von Frisch 1886–1982,* edited by B Hölldobler and M Lindauer, 91–107. Stuttgart: G. Fischer Verlag.

Franks NR. 1986. Teams in social insects: group retrieval of prey by army ants (*Eciton burchellii,* Hymenoptera: Formicidae). *Behavioral Ecology and Sociobiology* 18: 425–429.

Franks NR. 1989a. Army ants: a collective intelligence. *American Scientist* 77: 138–145.

Franks NR. 1989b. Thermoregulation in army ant bivouacs. *Physiological Entomology* 14: 397–404.

Franks NR. 2001. Evolution of mass transit systems in ants: a tale of two societies. In *Insect Movement: Mechanisms and Consequences,* edited by IP Woiwod, DR Reynolds, and CD Thomas, 281–298. Wallingford, United Kingdom: CAB International.

Franks NR, Bossert WH. 1983. The influence of swarm raiding army ants on the patchiness and diversity of a tropical leaf litter ant community. In *Tropical Rain Forest Ecology and Management,* edited by EL Sutton, TC Whitmore, and AC Chadwick, 151–163. Oxford: Blackwell.

Franks NR, Fletcher CR. 1983. Spatial patterns in army ant foraging and migration: *Eciton burchelli* on Barro Colorado Island, Panama. *Behavioral Ecology and Sociobiology* 12: 261–270.

Franks NR, Hölldobler B. 1987. Sexual competition during colony reproduction in army ants. *Biological Journal of the Linnean Society* 30: 229–243.

Franks NR, Gomez N, Goss S, Deneubourg JL. 1991. The blind leading the blind in army ant raid patterns: testing a model of self-organization (Hymenoptera: Formicidae). *Journal of Insect Behavior* 4: 583–607.

Franks NR, Sendova-Franks AB, Simmons J, Mogie M. 1999. Convergent evolution, superefficient teams and tempo in Old and New World army ants. *Proceedings of the Royal Society of London, Series B: Biological Sciences* 266: 1697–1701.

Franks NR, Sendova-Franks AB, Anderson C. 2001. Division of labour within teams of New World and Old World army ants. *Animal Behaviour* 62: 635–642.

Franks NR, Pratt SC, Mallon EB, Britton NF, Sumpter DJT. 2002. Information flow, opinion polling and collective intelligence in house-hunting social insects. *Philosophical Transactions of the Royal Society of London, Series B: Biological Sciences* 357: 1567–1583.

Freitas AVL. 1995. Nest relocation and prey specialization in the ant *Leptogenys propefalcigera* Roger (Formicidae: Ponerinae) in an urban area in southeastern Brazil. *Insectes Sociaux* 42: 453–456.

Funaro CF, Kronauer DJC, Moreau CS, Goldman-Huertas B, Pierce NE, Russell JA. 2011. Army ants harbor a host-specific clade of *Entomoplasmatales* bacteria. *Applied and Environmental Microbiology* 77: 346–350.

Gal A, Saragosti J, Kronauer DJC. 2020. anTraX: high throughput video tracking of color-tagged insects. *bioRxiv* 2020.04.29.068478. Available at https://doi.org/10.1101/2020.04.29.068478.

García Márquez G. 1970. *One Hundred Years of Solitude.* New York: Harper & Row.

Garnier S, Kronauer DJC. 2017. The adaptive significance of phasic colony cycles in army ants. *Journal of Theoretical Biology* 428: 43–47.

Garnier S, Murphy T, Lutz M, Hurme E, Leblanc S, Couzin ID. 2013. Stability and responsiveness in a self-organized living architecture. *PLoS Computational Biology* 9: e1002984.

Gehlbach FR, Watkins JF II, Reno HW. 1968. Blind snake defensive behavior elicited by ant attacks. *BioScience* 18: 784–785.

Getz WM, Brückner D, Parisian TR. 1982. Kin structure and the swarming behavior of the honey bee *Apis mellifera. Behavioral Ecology and Sociobiology* 10: 265–270.

Gillespie DS, Cole AC. 1950. Measurements in the worker caste of two species of *Eciton* (*Neivamyrmex* Borgmeier) (Hymenoptera: Formicidae). *American Midland Naturalist* 44: 203–204.

Gobin B, Rüppell O, Hartmann A, Jungnickel H, Morgan ED, Billen J. 2001. A new type of exocrine gland and its function in mass recruitment in the ant *Cylindromyrmex whymperi* (Formicidae: Cerapachyinae). *Naturwissenschaften* 88: 395–399.

Gochfeld M, Tudor G. 1978. Ant-following birds in South American subtropical forests. *Wilson Bulletin* 90: 139–141.

Gómez Durán JM. 2016. Notes on the copulatory biology of the genus *Leptanilla* (Hymenoptera: Formicidae: Leptanillinae). Available at https://doi.org/10.13140/RG.2.1.2662.7446.

Gotwald WH Jr. 1972. Analogous prey escape mechanisms in a pulmonate mollusk and lepidopterous larvae. *Journal of the New York Entomological Society* 80: 111–113.

Gotwald WH Jr. 1974. Predatory behavior and food preferences of driver ants in selected African habitats. *Annals of the Entomological Society of America* 67: 877–886.

Gotwald WH Jr. 1976. Behavioral observations on African army ants of the genus *Aenictus* (Hymenoptera: Formicidae). *Biotropica* 8: 59–65.

Gotwald WH Jr. 1978. Trophic ecology and adaptation in tropical Old World ants of the subfamily Dorylinae (Hymenoptera: Formicidae). *Biotropica* 10: 161–169.

Gotwald WH Jr. 1979. Phylogenetic implications of army ant zoogeography (Hymenoptera: Formicidae). *Annals of the Entomological Society of America* 72: 462–467.

Gotwald WH Jr. 1982. Army ants. In *Social Insects,* Vol. 4, edited by HR Hermann, 157–254. New York: Academic Press.

Gotwald WH Jr. 1984–1985. Death on the march. *Rotunda* 17: 37–41.

Gotwald WH Jr. 1985. Reflections on the evolution of army ants (Hymenoptera, Formicidae). *Actes des Colloques Insectes Sociaux* 2: 7–16.

Gotwald WH Jr. 1988. On becoming an army ant. In *Advances in Myrmecology,* edited by JC Trager, 227–235. Leiden: E. J. Brill.

Gotwald WH Jr. 1995. *Army Ants: The Biology of Social Predation.* Ithaca, NY: Cornell University Press.

Gotwald WH Jr. 1996. Mites that live with army ants: a natural history of some myrmecophilous hitch-hikers, browsers, and parasites. *Journal of the Kansas Entomological Society* 69: 232–237.

Gotwald WH Jr, Brown WL Jr. 1966. The ant genus *Simopelta* (Hymenoptera: Formicidae). *Psyche* 73: 261–277.

Gotwald WH Jr, Cunningham-van Someren GR. 1976. Taxonomic and behavioral notes on the African ant, *Aenictus eugenii* Emery, with a description of the queen (Hymenoptera: Formicidae). *Journal of the New York Entomological Society* 84: 182–188.

Gotwald WH Jr, Schaefer RF Jr. 1982. Taxonomic implications of doryline worker ant morphology: *Dorylus* subgenus *Anomma* (Hymenoptera: Formicidae). *Sociobiology* 7: 187–204.

Graham JM, Kao AB, Wilhelm DA, Garnier S. 2017. Optimal construction of army ant living bridges. *Journal of Theoretical Biology* 435: 184–198.

Gudger EW. 1925. Stitching wounds with the mandibles of ants and beetles. *Journal of the American Medical Association* 84: 1861–1864.

Guilding L. 1834. Observations on the work of Maria Sybilla Merian on the insects etc. of Surinam. *Magazine of Natural History and Journal of Zoology, Botany, Mineralogy, Geology and Meteorology* 7: 355–375.

Haelewaters D, Verhaeghen SJC, Ríos Gonzáles TA, Bernal Vega JA, Villarreal Saucedo RV. 2017. New and interesting Laboulbeniales from Panama and neighboring areas. *Nova Hedwigia* 105: 267–299.

Hagan HR. 1954a. The reproductive system of the army-ant queen, *Eciton* (*Eciton*). Part 1. General anatomy. *American Museum Novitates* 1663: 1–12.

Hagan HR. 1954b. The reproductive system of the army-ant queen, *Eciton* (*Eciton*). Part 2. Histology. *American Museum Novitates* 1664: 1–17.

Hagan HR. 1954c. The reproductive system of the army-ant queen, *Eciton* (*Eciton*). Part 3. The oöcyte cycle. *American Museum Novitates* 1665: 1–20.

Hamilton WD. 1964a. The genetical evolution of social behaviour. I. *Journal of Theoretical Biology* 7: 1–16.

Hamilton WD. 1964b. The genetical evolution of social behaviour. II. *Journal of Theoretical Biology* 7: 17–52.

Hamilton WD. 1975. Gamblers since life began: barnacles, aphids, elms. *Quarterly Review of Biology* 50: 175–180.

Hammond RL, Keller L. 2004. Conflict over male parentage in social insects. *PLoS Biology* 2: e248.

Harper LH. 1989. The persistence of ant-following birds in small Amazonian forest fragments. *Acta Amazonica* 19: 249–263.

Hashimoto Y, Yamane S. 2014. Comparison of foraging habits between four sympatric army ant species of the genus *Aenictus* in Sarawak, Borneo. *Asian Myrmecology* 6: 95–104.

Heape W. 1931. *Emigration, Migration, and Nomadism.* London: Heffer & Sons.

Heinze J. 2016. The male has done his work—the male may go. *Current Opinion in Insect Science* 16: 22–27.

Helava JVT, Howden HF, Ritchie AJ. 1985. A review of the New World genera of the myrmecophilous and termitophilous subfamily Hetaeriinae. Sociobiology 10: 127–386.

Hermann HR Jr. 1968. Group raiding in *Termitopone commutata* (Roger) (Hymenoptera: Formicidae). *Journal of the Georgia Entomological Society* 3: 23–24.

Hirosawa H, Higashi S, Mohamed M. 2000. Food habits of *Aenictus* army ants and their effects on the ant community in a rain forest of Borneo. *Insectes Sociaux* 47: 42–49.

Hlaváč P, Janda M. 2009. *Leptogenopapus mirabilis,* a new genus and species of Lomechusini (Coleoptera: Staphylinidae, Aleocharinae) from Papua New Guinea associated with ants of the genus *Leptogenys* Roger. *Zootaxa* 2062: 57–64.

Hoenle PO. 2019. *Leptogenys.* In *Encyclopedia of Social Insects,* edited by CK Starr. Cham: Springer.

Hoenle PO, Blüthgen N, Brückner A, Kronauer DJC, Fiala B, Donoso DA, Smith MA, Ospina Jara B, von Beeren C. 2019. Species-level predation network uncovers high prey specificity in a Neotropical army ant community. *Molecular Ecology* 28: 2423–2440.

Hojo MK, Pierce NE, Tsuji K. 2015. Lycaenid caterpillar secretions manipulate attendant ant behavior. *Current Biology* 25: 2260–2264.

Hölldobler B. 1970. Zur Physiologie der Gast-Wirt-Beziehung (Myrmecophilie) bei Ameisen. II. Das Gastverhältnis des imaginalen *Atemeles pubicollis* Bris. (Col. Staphylidae) zu *Myrmica* und *Formica* (Hym. Formicidae). *Zeitschrift für Vergleichende Physiologie* 66: 215–250.

Hölldobler B. 1982. Communication, raiding behaviour and prey storage in *Cerapachys* (Hymenoptera: Formicidae). *Psyche* 89: 3–23.

Hölldobler B. 1983. Territorial behavior in the green tree ant (*Oecophylla smaragdina*). *Biotropica* 15: 241–250.

Hölldobler B. 2016. Queen specific exocrine glands in legionary ants and their possible function in sexual selection. *PLoS One* 11: e0151604.

Hölldobler B, Traniello JFA. 1980. The pygidial gland and chemical recruitment communication in *Pachycondyla (= Termitopone) laevigata. Journal of Chemical Ecology* 6: 883–893.

Hölldobler B, Wilson EO. 1990. *The Ants.* Cambridge, MA: Belknap Press of Harvard University Press.

Hölldobler B, Wilson EO. 2009. *The Superorganism. The Beauty, Elegance, and Strangeness of Insect Societies.* New York: W. W. Norton.

Hölldobler B, Braun U, Gronenberg W, Kirchner WH, Peeters C. 1994a. Trail communication in the ant *Megaponera foetens* (Fabr.) (Formicidae, Ponerinae). *Journal of Insect Physiology* 40: 585–593.

Hölldobler B, Peeters C, Obermayer M. 1994b. Exocrine glands and the attractiveness of the ergatoid queen in the ponerine ant *Megaponera foetens. Insectes Sociaux* 41: 63–72.

Hölldobler B, Janssen E, Bestmann HJ, Leal IR, Oliveira PS, Kern F, König WA. 1996. Communication in the migratory termite-hunting ant *Pachycondyla (= Termitopone) marginata* (Formicidae, Ponerinae). *Journal of Comparative Physiology A* 178: 47–53.

Hollingsworth MJ. 1960. Studies on the polymorphic workers of the army ant *Dorylus* (*Anomma*) *nigricans* Illiger. *Insectes Sociaux* 7: 17–37.

Huang MH. 2010. Multi-phase defense by the big-headed ant, *Pheidole obtusospinosa,* against raiding army ants. *Journal of Insect Science* 10: 1.

Hughes WOH, Oldroyd BP, Beekman M, Ratnieks FLW. 2008. Ancestral monogamy shows kin selection is key to the evolution of eusociality. *Science* 320: 1213–1216.

Hughes DP, Pierce NE, Boomsma JJ. 2008. Social insect symbionts: evolution in homeostatic fortresses. *Trends in Ecology and Evolution* 23: 672–677.

Hurlbert AH, Ballantyne IVF, Powell S. 2008. Shaking a leg and hot to trot: the effects of body size and temperature on running speed in ants. *Ecological Entomology* 33: 144–154.

Huxley J. 1930. *Ants.* New York: Jonathan Cape & Harrison Smith.

Idogawa N, Dobata S. 2018. Colony structure and life history of *Lioponera daikoku* (Formicidae: Dorylinae). *Asian Myrmecology* 10: e010006.

Ito F, Ohkawara K. 2000. Production and behavior of ergatoid queens in two species of the Indonesian ponerine ant genus *Leptogenys* (*diminuta*-group) (Hymenoptera: Formicidae). *Annals of the Entomological Society of America* 93: 869–873.

Ito F, Jaitrong W, Hashim R, Mizuno R. 2018. Colony composition, brood production and caste dimorphism in two species of the doryline genus *Lioponera* in the Oriental tropics (Formicidae: Dorylinae). *Asian Myrmecology* 10: e010007.

Ivens ABF, von Beeren C, Blüthgen N, Kronauer DJC. 2016. Studying the complex communities of ants and their symbionts using ecological network analysis. *Annual Review of Entomology* 61: 353–371.

Jackson WB. 1957. Microclimatic patterns in the army ant bivouac. *Ecology* 38: 276–285.

Jaffé R, Kronauer DJC, Kraus FB, Boomsma JJ, Moritz RFA. 2007. Worker caste determination in the army ant *Eciton burchellii. Biology Letters* 3: 513–516.

Jaffé R, Moritz RFA, Kraus FB. 2009. Gene flow is maintained by polyandry and male dispersal in the army ant *Eciton burchellii. Population Ecology* 51: 227–236.

Jaitrong W, Ruangsittichai J. 2018. Two new species of the *Aenictus wroughtonii* species group (Hymenoptera, Formicidae, Dorylinae) from Thailand. *ZooKeys* 775: 103–115.

Janssen R, Witte V. 2003. *Allopeas myrmekophilos* n. sp., the first snail reported as living in army ant colonies. *Archiv für Molluskenkunde* 131: 211–215.

Jeanne RL. 1970. Chemical defense of brood by a social wasp. *Science* 168: 1465–1466.

Jeanne RL. 1975. The adaptiveness of social wasp nest architecture. *Quarterly Review of Biology* 50: 267–287.

Jolivet P. 1952. Quelques données sur la myrmécophilie des clytrides (Col. Chrysomeloidea). *Bulletin de I'lnstitut Royal des Sciences Naturelles de Belgique* 28: 1–12.

Jurine L. 1807. *Nouvelle méthode de classer les Hyménoptères et les Diptères. Hyménoptères.* Vol. 1. Geneva: Paschoud.

Kaspari M, O'Donnell S. 2003. High rates of army ant raids in the Neotropics and implications for ant colony and community structure. *Evolutionary Ecology Research* 5: 933–939.

Kaspari M, Powell S, Lattke J, O'Donnell S. 2011. Predation and patchiness in the tropical litter: do swarm-raiding army ants skim the cream or drain the bottle? *Journal of Animal Ecology* 80: 818–823.

Keegans SJ, Billen J, Morgan ED, Gökcen OA. 1993. Volatile glandular secretions of three species of New World army ants, *Eciton burchelli, Labidus coecus,* and *Labidus praedator. Journal of Chemical Ecology* 19: 2705–2719.

Kirby W, Spence W. 1843. *An Introduction to Entomology.* Vol. 2. 6th ed. London: Longman, Brown, Green, and Longmans.

Kistner DH. 1966a. A revision of the African species of the aleocharine tribe Dorylomimini (Coleoptera: Staphylinidae). II. The genera *Dorylomimus, Dorylonannus, Dorylogaster, Dorylobactrus,* and *Mimanomma,* with notes on their behavior. *Annals of the Entomological Society of America* 59: 320–340.

Kistner DH. 1966b. A revision of the myrmecophilous tribe Deremini (Coleoptera: Staphylinidae). Part I. The *Dorylopora* complex and their behavior. *Annals of the Entomological Society of America* 59: 341–358.

Kistner DH. 1968. Revision of the myrmecophilous species of the tribe Myrmedonini. Part II. The genera *Aenictonia* and *Anommatochara*—their relationships and behavior. *Annals of the Entomological Society of America* 61: 971–986.

Kistner DH. 1976. The natural history of the myrmecophilous tribe Pygostenini (Coleoptera: Staphylinidae). Section 3. Behavior and food habits of the Pygostenini. *Sociobiology* 2: 171–188.

Kistner DH. 1977. A revision of the myrmecophilous genus *Anepipleuronia* with a description of its behavior and epidermal glands. *Sociobiology* 2: 226–252.

Kistner DH. 1979. Social and evolutionary significance of social insect symbionts. In *Social Insects,* Vol. 1, edited by HR Hermann, 339–413. New York: Academic Press.

Kistner DH. 1982. The social insects' bestiary. In *Social Insects,* Vol. 3, edited by HR Hermann, 1–244. New York: Academic Press.

Kistner DH. 1983. A new genus and twelve new species of ant mimics associated with *Pheidologeton* (Coleoptera, Staphylinidae; Hymenoptera, Formicidae). *Sociobiology* 8: 155–198.

Kistner DH, Berghoff SM. 2006. First species of *Vatesus* captured with *Cheliomyrmex* (Coleoptera: Staphylinidae) (Hymenoptera: Formicidae). *Sociobiology* 48: 385–394.

Kistner DH, Davis LN. 1989. New species of *Notoxopria* and their behavior, with notes on *Mimopria* (Hymenoptera: Diapriidae). *Sociobiology* 16: 217–238.

Kistner DH, Jacobson HR. 1975. A review of the myrmecophilous Staphylinidae associated with *Aenictus* in Africa and the Orient (Coleoptera; Hymenoptera, Formicidae) with notes on their behavior and glands. *Sociobiology* 1: 20–73.

Kistner DH, Jacobson HR. 1990. Cladistic analysis and taxonomic revision of the ecitophilous tribe Ecitocharini with studies of their behavior and evolution (Coleoptera, Staphylinidae, Aleocharinae). *Sociobiology* 17: 333–480.

Kistner DH, Mooney RR. 2011. Revision of the genus *Ecitopora* with descriptions of new species (Coleoptera: Staphylinidae). *Sociobiology* 58: 269–308.

Kistner DH, von Beeren C, Witte V. 2008. Redescription of the generitype of *Trachydonia* and a new host record for *Maschwitzia ulrichi* (Coleoptera: Staphylinidae). *Sociobiology* 52: 497–524.

Kley NJ. 2001. Prey transport mechanisms in blindsnakes and the evolution of unilateral feeding systems in snakes. *American Zoologist* 41: 1321–1337.

Koblick TA, Kistner DH. 1965. A revision of the species of the genus *Myrmechusa* from tropical Africa with notes on their behavior and their relationship to the Pygostenini (Coleoptera: Staphylinidae). *Annals of the Entomological Society of America* 58: 28–44.

Koops K, Schöning C, Isaji M, Hashimoto C. 2015a. Cultural differences in ant-dipping tool length between neighbouring chimpanzee communities at Kalinzu, Uganda. *Scientific Reports* 5: 12456.

Koops K, Schöning C, McGrew WC, Matsuzawa T. 2015b. Chimpanzees prey on army ants at Seringbara, Nimba Mountains, Guinea: predation patterns and tool use characteristics. American Journal of Primatology 77: 319–329.

Korb J, Heinze J. 2008. The ecology of social life: a synthesis. In *Ecology of Social Evolution,* edited by J Korb and J Heinze, 245–259. Berlin: Springer.

Korb J, Heinze J. 2016. Major hurdles for the evolution of sociality. *Annual Review of Entomology* 61: 297–316.

Krell F-T. 1999. Dung beetles unharmed by army ants in tropical Africa (Coleoptera: Scarabaeidae; Hymenoptera: Formicidae, Dorylinae). *Coleopterists Bulletin* 53: 325–328.

Kroll JC, Gehlbach FR, Watkins JF II. 1971. Pheromone trail-following studies of typhlopid, leptotyphlopid, and colubrid snakes. *Behaviour* 40: 282–294.

Kronauer DJC. 2009. Recent advances in army ant biology (Hymenoptera: Formicidae). *Myrmecological News* 12: 51–65.

Kronauer DJC. 2020. Army ants. In *Encyclopedia of Social Insects,* edited by CK Starr. Cham: Springer.

Kronauer DJC, Boomsma JJ. 2007a. Do army ant queens re-mate later in life? *Insectes Sociaux* 54: 20–28.

Kronauer DJC, Boomsma JJ. 2007b. Multiple queens means fewer mates. *Current Biology* 17: R753–R755.

Kronauer DJC, Libbrecht R. 2018. Back to the roots: the importance of using simple insect societies to understand the molecular basis of complex social life. *Current Opinion in Insect Science* 28: 33–39.

Kronauer DJC, Pierce NE. 2011. Myrmecophiles. *Current Biology* 21: R208–R209.

Kronauer DJC, Schöning C, Pedersen JS, Boomsma JJ, Gadau J. 2004. Extreme queen-mating frequency and colony fission in African army ants. *Molecular Ecology* 13: 2381–2388.

Kronauer DJC, Berghoff SM, Powell S, Denny AJ, Edwards KJ, Franks NR, Boomsma JJ. 2006a. A reassessment of the mating system characteristics of the army ant *Eciton burchellii*. *Naturwissenschaften* 93: 402–406.

Kronauer DJC, Schöning C, Boomsma JJ. 2006b. Male parentage in army ants. *Molecular Ecology* 15: 1147–1151.

Kronauer DJC, Johnson RA, Boomsma JJ. 2007a. The evolution of multiple mating in army ants. *Evolution* 61: 413–422.

Kronauer DJC, Rodríguez Ponce ER, Lattke JE, Boomsma JJ. 2007b. Six weeks in the life of a reproducing army ant colony: male parentage and colony behaviour. *Insectes Sociaux* 54: 118–123.

Kronauer DJC, Schöning C, Vilhelmsen LB, Boomsma JJ. 2007c. A molecular phylogeny of *Dorylus* army ants provides evidence for multiple evolutionary transitions in foraging niche. *BMC Evolutionary Biology* 7: 56.

Kronauer DJC, Schöning C, d'Ettorre P, Boomsma JJ. 2010. Colony fusion and worker reproduction after queen loss in army ants. *Proceedings of the Royal Society of London, Series B: Biological Sciences* 277: 755–763.

Kronauer DJC, O'Donnell S, Boomsma JJ, Pierce NE. 2011a. Strict monandry in the ponerine army ant genus *Simopelta* suggests that colony size and complexity drive mating system evolution in social insects. *Molecular Ecology* 20: 420–428.

Kronauer DJC, Peters MK, Schöning C, Boomsma JJ. 2011b. Hybridization in East African swarm-raiding army ants. *Frontiers in Zoology* 8: 20.

Kronauer DJC, Pierce NE, Keller L. 2012. Asexual reproduction in introduced and native populations of the ant *Cerapachys biroi. Molecular Ecology* 21: 5221–5235.

Kronauer DJC, Tsuji K, Pierce NE, Keller L. 2013. Non-nest mate discrimination and clonal colony structure in the parthenogenetic ant *Cerapachys biroi. Behavioral Ecology* 24: 617–622.

Kumar A, O'Donnell S. 2007. Fragmentation and elevation effects on bird-army ant interactions in neotropical montane forest of Costa Rica. *Journal of Tropical Ecology* 23: 581–590.

Kumar A, O'Donnell S. 2009. Elevation and forest clearing effects on foraging differ between surface- and subterranean-foraging army ants (Formicidae: Ecitoninae). *Journal of Animal Ecology* 78: 91–97.

Lalor PF, Hughes WOH. 2011. Alarm behaviour in *Eciton* army ants. *Physiological Entomology* 36: 1–7.

LaMon B, Topoff H. 1981. Avoiding predation by army ants: defensive behaviours of three ant species of the genus *Camponotus. Animal Behaviour* 29: 1070–1081.

Lanan M. 2014. Spatiotemporal resource distribution and foraging strategies of ants (Hymenoptera: Formicidae). *Myrmecological News* 20: 53–70.

LaPolla JS, Mueller UG, Seid M, Cover SP. 2002. Predation by the army ant *Neivamyrmex rugulosus* on the fungus-growing ant *Trachymyrmex arizonensis. Insectes Sociaux* 49: 251–256.

LaPolla JS, Dlussky GM, Perrichot V. 2013. Ants and the fossil record. *Annual Review of Entomology* 58: 609–630.

Lappano ER. 1958. A morphological study of larval development in polymorphic all-worker broods of the army ant *Eciton burchelli. Insectes Sociaux* 5: 31–66.

Latreille PA. 1802. *Histoire Naturelle des Fourmis, et Recueil de Mémoires et d'Observations sur les Abeilles, les Araignées, les Faucheurs, et Autres Insectes.* Paris: Crapelet.

Latreille PA. 1804. Tableau méthodique des insectes. In *Nouveau Dictionnaire d'Histoire Naturelle.* Tome 24, 129–200. Paris: Déterville.

Lattke JE. 2011. Revision of the New World species of the genus *Leptogenys* Roger (Insecta: Hymenoptera: Formicidae: Ponerinae). *Arthropod Systematics and Phylogeny* 69: 127–264.

Leakey LSB. 1974. *By the Evidence: Memoirs, 1932–1951.* New York: Harcourt Brace Jovanovich.

Leal IR, Oliveira PS. 1995. Behavioral ecology of the neotropical termite-hunting ant *Pachycondyla* (= *Termitopone*) *marginata:* colony founding, group-raiding and migratory patterns. *Behavioral Ecology and Sociobiology* 37: 373–383.

Le Breton J, Dejean A, Snelling G, Orivel J. 2007. Specialized predation on *Wasmannia auropunctata* by the army ant species *Neivamyrmex compressinodis. Journal of Applied Entomology* 131: 740–743.

Le Conte Y, Arnold G, Trouiller J, Masson C, Chappe B. 1990. Identification of a brood pheromone in honeybees. *Naturwissenschaften* 77: 334–336.

Lenoir A, D'Ettorre P, Errard C, Hefetz A. 2001. Chemical ecology and social parasitism in ants. *Annual Review of Entomology* 46: 573–599.

Lenoir A, Chalon Q, Carvajal A, Ruel C, Barroso Á, Lackner T, Boulay R. 2012. Chemical integration of myrmecophilous guests in *Aphaenogaster* ant nests. *Psyche* 2012: 840860.

Leroux JM. 1977a. Formation et déroulement des raids de chasse d'*Anomma nigricans* Illiger (Hym. Dorylinae) dans une savane de Côte d'Ivoire. *Bulletin de la Société Zoologique de France* 102: 445–458.

Leroux JM. 1977b. Densité des colonies et observations sur les nids des dorylines *Anomma nigricans* Illiger (Hym. Formicidae) dans la région de Lamto (Côte d'Ivoire). *Bulletin de la Société Zoologique de France* 102: 51–62.

Leroux JM. 1979a. Possibilités de scissions multiples pour des colonies de dorylines *Anomma nigricans* Illiger Hyménoptères Formicidae, en Côte d'Ivoire. *Insectes Sociaux* 26: 13–17.

Leroux JM. 1979b. Sur quelques modalités de disparition des colonies d'*Anomma nigricans* Illiger (Formicidae Dorylinae) dans la région de Lamto (Côte d'Ivoire). *Insectes Sociaux* 26: 93–100.

Leroux JM. 1982. Ecologie des populations de dorylines *Anomma nigricans* Illiger (Hym. Formicidae) dans la région de Lamto (Côte d'Ivoire). Publications du Laboratoire de Zoologie, no. 22. Paris: Ecole Normale Supérieure.

Leston D. 1979. Dispersal by male doryline ants in West Africa. *Psyche* 86: 63–77.

Libbrecht R, Oxley PR, Kronauer DJC. 2018. Clonal raider ant brain transcriptomics identifies candidate molecular mechanisms for reproductive division of labor. *BMC Biology* 16: 89.

Linnaeus C. 1758. *Systema Naturae.* Stockholm: L. Salvius.

Liu C, Hita Garcia F, Peng Y-Q, Economo EP. 2015. *Aenictus yangi* sp. n.—a new species of the *A. ceylonicus* species group (Hymenoptera, Formicidae, Dorylinae) from Yunnan, China. *Journal of Hymenoptera Research* 42: 33–45.

Logan CJ, O'Donnell S, Clayton NS. 2011. A case of mental time travel in ant-following birds? *Behavioral Ecology* 22: 1149–1153.

Loiácono MS, Margaría CB, Aquino DA. 2013. Diapriinae wasps (Hymenoptera: Diaprioidea: Diapriidae) associated with ants (Hymenoptera: Formicidae) in Argentina. *Psyche* 2013: 320590.

Longhurst C, Howse PE. 1979. Foraging, recruitment and emigration in *Megaponera foetens* (Fab.) (Hymenoptera: Formicidae) from the Nigerian Guinea Savanna. *Insectes Sociaux* 26: 204–215.

Longhurst C, Baker R, Howse PE. 1979a. Termite predation by *Megaponera foetens* (Fab.) (Hymenoptera: Formicidae). Coordination of raids by glandular secretions. *Journal of Chemical Ecology* 5: 703–719.

Longhurst C, Johnson RA, Wood TG. 1979b. Foraging, recruitment and predation by *Decamorium uelense* (Santschi) (Formicidae: Myrmicinae) on termites in southern Guinea Savanna, Nigeria. *Oecologia* 38: 83–91.

Longino JT. 2005. Complex nesting behavior by two Neotropical species of the ant genus *Stenamma* (Hymenoptera: Formicidae). *Biotropica* 37: 670–675.

Longino JT, Branstetter MG. n.d. A trove of New World *Syscia* (Formicidae, Dorylinae). Unpublished manuscript.

Longino JT, Coddington J, Colwell RK. 2002. The ant fauna of a tropical rain forest: estimating species richness three different ways. *Ecology* 83: 689–702.

Lovejoy TE, Bierregaard RO, Rylands AB, Malcolm JR, Quintela CE, Harper LH, Brown KS, Powell AH, Powell GVN, Schubart HOR, Hays MB. 1986. Edge and other effects of isolation on Amazon forest fragments. In *Conservation Biology: The Science of Scarcity and Diversity,* edited by ME Soule, 257–285. Sunderland, MA: Sinauer Associates.

Loveridge A. 1922. Account of an invasion of "siafu" or red driver-ants—*Dorylus* (*Anomma*) *nigricans* Illig. *Proceedings of the Royal Entomological Society of London* 5: 33–46.

Loveridge A. 1944. *Many Happy Days I've Squandered*. New York: Harper & Brothers.

Lucky A, Trautwein MD, Guénard B, Weiser MD, Dunn RR. 2013. Tracing the rise of the ants—out of the ground. *PLoS One* 8: e84012.

Łukasik P, Newton JA, Sanders JG, Hu Y, Moreau CS, Kronauer DJC, O'Donnell S, Koga R, Russell JA. 2017. The structured diversity of specialized gut symbionts of the New World army ants. *Molecular Ecology* 26: 3808–3825.

Macevicz S. 1979. Some consequences of Fisher's sex ratio principle for social Hymenoptera that reproduce by colony fission. *American Naturalist* 113: 363–371.

Mackay WP. 1996. A revision of the ant genus *Acanthostichus* (Hymenoptera: Formicidae). *Sociobiology* 27: 129–179.

Mackay WP, Mackay EE. 2002. *The Ants of New Mexico*. Lewiston: Edwin Mellen Press.

Mackay WP, Mackay EE. 2008. Revision of the ants of the genus *Simopelta* Mann. In *Sistemática, Biogeografía y Conservación de las Hormigas Cazadoras de Colombia,* edited by E Jiménez, F Fernández, TM Arias, FH Lozano-Zambrano, 285–328. Bogotá: Instituto de Investigación de Recursos Biológicos Alexander von Humboldt.

Maisonnasse A, Lenoir J-C, Beslay D, Crauser D, Le Conte Y. 2010. E-β-ocimene, a volatile brood pheromone involved in social regulation in the honey bee colony (*Apis mellifera*). *PLoS One* 5: e13531.

Manubay JA, Powell S. 2020. Detection of prey odors underpins dietary specialization in a Neotropical top-predator: how army ants find their ant prey. *Journal of Animal Ecology* 89: 1165–1174.

Mariano CSF, Delabie JHC, Pompolo SG. 2004. Nota sobre uma colônia e o cariótipo da formiga neotropical *Cylindromyrmex brasiliensis* Emery (Hymenoptera: Formicidae: Cerapachyinae). *Neotropical Entomology* 33: 267–269.

Martínez AE, Pollock HS, Kelley JP, Tarwater CE. 2018. Social information cascades influence the formation of mixed-species foraging aggregations of ant-following birds in the Neotropics. *Animal Behaviour* 135: 25–35.

Martins MM. 2000. Foraging over army ants by *Callithrix aurita* (Primates: Callitrichidae): seasonal occurrence? *Revista de Biología Tropical* 48: 261–262.

Maruyama M, Disney RHL. 2008. Scuttle flies associated with Old World army ants in Malaysia (Diptera: Phoridae; Hymenoptera, Formicidae, Dorylinae). *Sociobiology* 51: 65–71.

Maruyama M, Parker J. 2017. Deep-time convergence in rove beetle symbionts of army ants. *Current Biology* 27: 920–926.

Maruyama M, Disney RHL, Hashim R. 2008. Three new species of legless, wingless scuttle flies (Diptera: Phoridae) associated with army ants (Hymenoptera: Formicidae) in Malaysia. *Sociobiology* 52: 485–496.

Maruyama M, Akino T, Hashim R, Komatsu T. 2009. Behavior and cuticular hydrocarbons of myrmecophilous insects (Coleoptera: Staphylinidae; Diptera: Phoridae; Thysanura) associated with Asian *Aenictus* army ants (Hymenoptera; Formicidae). *Sociobiology* 54: 19–35.

Maruyama M, von Beeren C, Hashim R. 2010. Aleocharine rove beetles (Coleoptera, Staphylinidae) associated with *Leptogenys* Roger, 1861 (Hymenoptera, Formicidae) I. Review of three genera associated with *L. distinguenda* (Emery, 1887) and *L. mutabilis* (Smith, 1861). *ZooKeys* 59: 47–60.

Maruyama M, Matsumoto T, Itioka T. 2011. Rove beetles (Coleoptera: Staphylinidae) associated with *Aenictus laeviceps* (Hymenoptera: Formicidae) in Sarawak, Malaysia: strict host specificity, and first myrmecoid Aleocharini. *Zootaxa* 3102: 1–26.

Maschwitz U, Hänel H. 1985. The migrating herdsman *Dolichoderus* (*Diabolus*) *cuspidatus*: an ant with a novel mode of life. *Behavioral Ecology and Sociobiology* 17: 171–184.

Maschwitz U, Mühlenberg M. 1975. Zur Jagdstrategie einiger orientalischer *Leptogenys*-Arten (Formicidae: Ponerinae). *Oecologia* 20: 65–83.

Maschwitz U, Schönegge P. 1983. Forage communication, nest moving recruitment, and prey specialization in the oriental ponerine *Leptogenys chinensis. Oecologia* 57: 175–182.

Maschwitz U, Steghaus-Kovac S. 1991. Individualismus versus Kooperation—Gegensätzliche Jagd- und Rekrutierungsstrategien bei tropischen Ponerinen (Hymenoptera: Formicidae). *Naturwissenschaften* 78: 103–113.

Maschwitz U, Steghaus-Kovac S, Gaube R, Hänel H. 1989. A South East Asian ponerine ant of the genus *Leptogenys* (Hym., Form.) with army ant life habits. *Behavioral Ecology and Sociobiology* 24: 305–316.

Masuko K. 1989. Larval hemolymph feeding in the ant *Leptanilla japonica* by use of a specialized duct organ, the "larval hemolymph tap" (Hymenoptera: Formicidae). *Behavioral Ecology and Sociobiology* 24: 127–132.

Masuko K. 1990. Behavior and ecology of the enigmatic ant *Leptanilla japonica* Baroni Urbani (Hymenoptera: Formicidae: Leptanillinae). *Insectes Sociaux* 37: 31–57.

Masuko K. 2006. Collection and the result of dissection of the ant *Cerapachys humicola. Ari: Journal of the Myrmecological Society of Japan* 28: 1–6.

Masuko K. 2019. Leptanillinae. In *Encyclopedia of Social Insects*, edited by CK Starr. Cham: Springer.

Maynard Smith J, Szathmáry E. 1995. *The Major Transitions in Evolution*. Oxford: Oxford University Press.

Mayr G. 1886. Ueber *Eciton-Labidus. Wiener Entomologische Zeitung* 5: 115–122.

McDonald P, Topoff H. 1986. The development of defensive behavior against predation by army ants. *Developmental Psychobiology* 19: 351–367.

McGlynn TP. 2012. The ecology of nest movement in social insects. *Annual Review of Entomology* 57: 291–308.

McGlynn TP, Carr RA, Carson JH, Buma J. 2004. Frequent nest relocation in the ant *Aphaenogaster araneoides:* resources, competition, and natural enemies. *Oikos* 106: 612–621.

McIver JD, Stonedahl G. 1993. Myrmecomorphy: morphological and behavioral mimicry of ants. *Annual Review of Entomology* 38: 351–377.

McKenzie SK, Kronauer DJC. 2018. The genomic architecture and molecular evolution of ant odorant receptors. *Genome Research* 28: 1757–1765.

McKenzie SK, Fetter-Pruneda I, Ruta V, Kronauer DJC. 2016. Transcriptomics and neuroanatomy of the clonal raider ant implicate an expanded clade of odorant receptors in chemical communication. *Proceedings of the National Academy of Sciences of the United States of America* 113: 14091–14096.

McKenzie SK, Winston ME, Grewe F, Vargas-Asensio G, Rodríguez N, Rubin BER, Murillo C, von Beeren C, Moreau CS, Suen G, Pinto-Tomás AA, Kronauer DJC. n.d. The genomic basis of army ant chemosensory adaptations. Unpublished manuscript.

Meisel JE. 2006. Thermal ecology of the Neotropical army ant *Eciton burchellii. Ecological Applications* 16: 913–922.

Mendes LF, von Beeren C, Witte V. 2011. *Malayatelura ponerophila*—a new genus and species of silverfish (Zygentoma, Insecta) from Malaysia, living in *Leptogenys* army-ant colonies (Formicidae). *Deutsche Entomologische Zeitschrift* 58: 193–200.

Merian MS. 1705. *Metamorphosis Insectorum Surinamensium*. Amsterdam: self-published.

Mill AE. 1982. Emigration of a colony of the giant termite hunter, *Pachycondyla commutata* (Roger) (Hymenoptera: Formicidae). *Entomologist's Monthly Magazine* 118: 243–245.

Mill AE. 1984. Predation by the ponerine ant *Pachycondyla commutata* on termites of the genus *Syntermes* in Amazonian rain forest. *Journal of Natural History* 18: 405–410.

Mirenda JT, Topoff H. 1980. Nomadic behavior of army ants in a desert-grassland habitat. *Behavioral Ecology and Sociobiology* 7: 129–135.

Mirenda JT, Eakins DG, Gravelle K, Topoff H. 1980. Predatory behavior and prey selection by army ants in a desert-grassland habitat. *Behavioral Ecology and Sociobiology* 7: 119–127.

Miyata H, Shimamura T, Hirosawa H, Higashi S. 2003. Morphology and phenology of the primitive ponerine army ant *Onychomyrmex hedleyi* (Hymenoptera: Formicidae: Ponerinae) in a highland rainforest of Australia. *Journal of Natural History* 37: 115–125.

Miyata H, Hirata M, Azuma N, Murakami T, Higashi S. 2009. Army ant behaviour in the poneromorph hunting ant *Onychomyrmex hedleyi* Emery (Hymenoptera: Formicidae; Amblyoponinae). *Australian Journal of Entomology* 48: 47–52.

Mizuno R, Suttiprapan P, Jaitrong W, Ito F. 2019. Daily and seasonal foraging activity of the Oriental non-army ant doryline *Cerapachys sulcinodis* complex (Hymenoptera: Formicidae). *Sociobiology* 66: 239–246.

Moffett MW. 1984. Swarm raiding in a myrmicine ant. *Naturwissenschaften* 71: 588–590.

Moffett MW. 1988a. Foraging behavior in the Malayan swarm-raiding ant *Pheidologeton silenus* (Hymenoptera: Formicidae: Myrmicinae). *Annals of the Entomological Society of America* 81: 356–361.

Moffett MW. 1988b. Foraging dynamics in the group-hunting myrmicine ant, *Pheidologeton diversus*. *Journal of Insect Behavior* 1: 309–331.

Moffett MW. 1988c. Nesting, emigrations, and colony foundation in two group-hunting myrmicine ants (Hymenoptera: Formicidae: *Pheidologeton*). In *Advances in Myrmecology*, edited by TC Trager, 355–370. Leiden: E. J. Brill.

Moffett MW. 2010. *Adventures among Ants*. Berkeley: University of California Press.

Moffett MW. 2019. Marauder ants (*Carebara* in part). In *Encyclopedia of Social Insects*, edited by CK Starr. Cham: Springer.

Monteiro AFM, Sujii ER, Morais HC. 2008. Chemically based interactions and nutritional ecology of *Labidus praedator* (Formicidae: Ecitoninae) in an agroecosystem adjacent to a gallery forest. *Revista Brasileira de Zoologia* 25: 674–681.

Moreau CS, Bell CD. 2013. Testing the museum versus cradle tropical biological diversity hypothesis: phylogeny, diversification, and ancestral biogeographic range evolution of the ants. *Evolution* 67: 2240–2257.

Moreau CS, Bell CD, Vila R, Archibald SB, Pierce NE. 2006. Phylogeny of the ants: diversification in the age of angiosperms. *Science* 312: 101–104.

Morel L, Vander Meer RK. 1988. Do ant brood pheromones exist? *Annals of the Entomological Society of America* 81: 705–710.

Mori A, Grasso DA, Visicchio R, Le Moli F. 2001. Comparison of reproductive strategies and raiding behaviour in facultative and obligatory slave-making ants: the case of *Formica sanguinea* and *Polyergus rufescens*. *Insectes Sociaux* 48: 302–314.

Müller F. 1879. *Ituna* and *Thyridia*: a remarkable case of mimicry in butterflies. *Proceedings of the Entomological Society of London* 1879: xx—xxix.

Müller W. 1886. Beobachtungen an Wanderameisen (*Eciton hamatum* Fabr.). *Kosmos* 1: 81–93.

Mynhardt G. 2013. Declassifying myrmecophily in the Coleoptera to promote the study of ant-beetle symbioses. *Psyche* 2013: 696401.

Niu YF, Feng YL, Xie JL, Luo FC. 2010. Noxious invasive *Eupatorium adenophorum* may be a moving target: implications of the finding of a native natural enemy, *Dorylus orientalis*. *Chinese Science Bulletin* 55: 3743–3745.

Ocampo-Ariza C, Kupsch D, Motombi FN, Bobo KS, Kreft H, Waltert M. 2019. Extinction thresholds and negative responses of Afrotropical ant-following birds to forest cover loss in oil palm and agroforestry landscapes. *Basic and Applied Ecology* 39: 26–37.

O'Donnell S. 2017. Evidence for facilitation among avian army-ant attendants: specialization and species associations across elevations. *Biotropica* 49: 665–674.

O'Donnell S, Jeanne RL. 1990. Notes on an army ant (*Eciton burchelli*) raid on a social wasp colony (*Agelaia yepocapa*) in Costa Rica. *Journal of Tropical Ecology* 6: 507–509.

O'Donnell S, Kumar A. 2006. Microclimatic factors associated with elevational changes in army ant density in tropical montane forest. *Ecological Entomology* 31: 491–498.

O'Donnell S, Kaspari M, Lattke J. 2005. Extraordinary predation by the Neotropical army ant *Cheliomyrmex andicola:* implications for the evolution of the army ant syndrome. *Biotropica* 37: 706–709.

O'Donnell S, Lattke J, Powell S, Kaspari M. 2007. Army ants in four forests: geographic variation in raid rates and species composition. *Journal of Animal Ecology* 76: 580–589.

O'Donnell S, Lattke J, Powell S, Kaspari M. 2009. Species and site differences in Neotropical army ant emigration behaviour. *Ecological Entomology* 34: 476–482.

O'Donnell S, Kumar A, Logan C. 2010. Army ant raid attendance and bivouac-checking behavior by Neotropical montane forest birds. *Wilson Journal of Ornithology* 122: 503–512.

O'Donnell S, Kaspari M, Kumar A, Lattke J, Powell S. 2011. Elevational and geographic variation in army ant swarm raid rates. *Insectes Sociaux* 58: 293–298.

O'Donnell S, Logan CJ, Clayton NS. 2012. Specializations of birds that attend army ant raids: an ecological approach to cognitive and behavioral studies. *Behavioural Processes* 91: 267–274.

O'Donnell S, Kumar A, Logan CJ. 2014. Do Nearctic migrant birds compete with residents at army ant raids? A geographic and seasonal analysis. *Wilson Journal of Ornithology* 126: 474–487.

O'Donnell S, Bulova S, Barrett M, von Beeren C. 2018. Brain investment under colony-level selection: soldier specialization in *Eciton* army ants (Formicidae: Dorylinae). *BMC Zoology* 3: 3.

Okabe K. 2013. Ecological characteristics of insects that affect symbiotic relationships with mites. *Entomological Science* 16: 363–378.

Otis GW, Santana E, Crawford DL, Higgins ML. 1986. The effect of foraging army ants on leaf-litter arthropods. *Biotropica* 18: 56–61.

Ott R, von Beeren C, Hashim R, Witte V, Harvey MS. 2015. *Sicariomorpha*, a new myrmecophilous goblin spider genus (Araneae, Oonopidae) associated with Asian army ants. *American Museum Novitates* 3843: 1–14.

Oxley PR, Ji L, Fetter-Pruneda I, McKenzie SK, Li C, Hu H, Zhang G, Kronauer DJC. 2014. The genome of the clonal raider ant *Cerapachys biroi. Current Biology* 24: 451–458.

Päivinen J, Ahlroth P, Kaitala V, Kotiaho JS, Suhonen J, Virola T. 2003. Species richness and regional distribution of myrmecophilous beetles. *Oecologia* 134: 587–595.

Palmer KA, Oldroyd BP. 2000. Evolution of multiple mating in the genus *Apis. Apidologie* 31: 235–248.

Pamilo P. 1991. Evolution of colony characteristics in social insects. I. Sex allocation. *American Naturalist* 137: 83–107.

Parker J. 2016. Myrmecophily in beetles (Coleoptera): evolutionary patterns and biological mechanisms. *Myrmecological News* 22: 65–108.

Parker J, Grimaldi DA. 2014. Specialized myrmecophily at the ecological dawn of modern ants. *Current Biology* 24: 2428–2434.

Parmentier T, Dekoninck W, Wenseleers T. 2016. Do well-integrated species of an inquiline community have a lower brood predation tendency? A test using red wood ant myrmecophiles. *BMC Evolutionary Biology* 16: 12.

Partridge LW, Britton NF, Franks NR. 1996. Army ant population dynamics: the effects of habitat quality and reserve size on population size and time to extinction. *Proceedings of the Royal Society of London, Series B: Biological Sciences* 263: 735–741.

Paulian R. 1948. Observations sur les coléoptères commensaux d'*Anomma nigricans* en Côte d'Ivoire. *Annales des Sciences Naturelles: Zoologie et Biologie Animale* 10: 79–102.

Peeters C, De Greef S. 2015. Predation on large millipedes and self-assembling chains in *Leptogenys* ants from Cambodia. *Insectes Sociaux* 62: 471–477.

Peeters C, Ito F. 2015. Wingless and dwarf workers underlie the ecological success of ants (Hymenoptera: Formicidae). *Myrmecological News* 21: 117–130.

Pérez-Espona S, McLeod JE, Franks NR. 2012. Landscape genetics of a top neotropical predator. *Molecular Ecology* 21: 5969–5985.

Pérez-Espona S, Goodall-Copestake WP, Berghoff SM, Edwards KJ, Franks NR. 2018. Army imposters: diversification of army ant-mimicking beetles with their *Eciton* hosts. *Insectes Sociaux* 65: 59–75.

Perfecto I. 1992. Observations of a *Labidus coecus* (Latreille) underground raid in the central highlands of Costa Rica. *Psyche* 99: 214–220.

Peters MK. 2010. Ant-following and the prevalence of blood parasites in birds of African rainforests. *Journal of Avian Biology* 41: 105–110.

Peters MK, Okalo B. 2009. Severe declines of ant-following birds in African rainforest fragments are facilitated by a subtle change in army ant communities. *Biological Conservation* 142: 2050–2058.

Peters MK, Likare S, Kraemer M. 2008. Effects of habitat fragmentation and degradation on flocks of African ant-following birds. *Ecological Applications* 18: 847–858.

Peters MK, Fischer G, Schaab G, Kraemer M. 2009. Species compensation maintains abundance and raid rates of African swarm-raiding army ants in rainforest fragments. *Biological Conservation* 142: 668–675.

Peters MK, Lung T, Schaab G, Wägele JW. 2011. Deforestation and the population decline of the army ant *Dorylus wilverthi* in western Kenya over the last century. *Journal of Applied Ecology* 48: 697–705.

Peters MK, Fischer G, Hita Garcia F, Lung T, Wägele JW. 2013. Spatial variation in army ant swarm raiding and its potential effect on biodiversity. *Biotropica* 45: 54–62.

Poinar GO Jr, Poinar R. 1999. *The Amber Forest. A Reconstruction of a Vanished World.* Princeton, NJ: Princeton University Press.

Poinar GJ, Lachaud JP, Castillo A, Infante F. 2006. Recent and fossil nematode parasites (Nematoda: Mermithidae) of Neotropical ants. *Journal of Invertebrate Pathology* 91: 19–26.

Pollock HS, Martínez AE, Kelley JP, Touchton JM, Tarwater CE. 2017. Heterospecific eavesdropping in ant-following birds of the Neotropics is a learned behaviour. *Proceedings of the Royal Society of London, Series B: Biological Sciences* 284: 20171785.

Powell S. 2011. How much do army ants eat? On the prey intake of a neotropical top-predator. *Insectes Sociaux* 58: 317–324.

Powell S, Baker B. 2008. Os grandes predadores dos neotrópicos: comportamento, dieta e impacto das formigas de correição (Ecitoninae). In *Insetos Sociais da Biologia à Aplicação,* edited by F Vilela, IA dos Santos, JE Serrão, JH Schoereder, J Lino-Neto, LA de O. Campos, 18–37. Viçosa, Brazil: Universidade Federal de Viçosa Press.

Powell S, Clark E. 2004. Combat between large derived societies: a subterranean army ant established as a predator of mature leaf-cutting ant colonies. *Insectes Sociaux* 51: 342–351.

Powell S, Franks NR. 2005. Caste evolution and ecology: a special worker for novel prey. *Proceedings of the Royal Society of London, Series B: Biological Sciences* 272: 2173–80.

Powell S, Franks NR. 2006. Ecology and the evolution of worker morphological diversity: a comparative analysis with *Eciton* army ants. *Functional Ecology* 20: 1105–1114.

Powell S, Franks NR. 2007. How a few help all: living pothole plugs speed prey delivery in the army ant *Eciton burchellii*. *Animal Behaviour* 73: 1067–1076.

Price PW. 1980. *Evolutionary Biology of Parasites.* Princeton, NJ: Princeton University Press.

Punzo F. 1974. Comparative analysis of the feeding habits of two species of Arizona blind snakes, *Leptotyphlops h. humilis* and *Leptotyphlops d. dulcis*. *Journal of Herpetology* 8: 153–156.

Queller D, Strassmann J. 1998. Kin selection and social insects. *BioScience* 48: 165–175.

Quevillon LE, Hughes DP. 2018. Pathogens, parasites, and parasitoids of ants: a synthesis of parasite biodiversity and epidemiological traits. *bioRxiv* 384495. Available at https//doi. org/10.1101/384495.

Quiroga H. 1981. *Cuentos.* Caracas: Biblioteca Ayacucho.

Rabeling C, Brown JM, Verhaagh M. 2008. Newly discovered sister lineage sheds light on early ant evolution. *Proceedings of the National Academy of Sciences of the United States of America* 105: 14913–14917.

Raignier A. 1972. Sur l'origine des nouvelles sociétés des fourmis voyageuses africaines (Hyménoptères, Formicidae, Dorylinae). *Insectes Sociaux* 19: 153–170.

Raignier A, van Boven JKA. 1955. Etude taxonomique, biologique et biométrique des *Dorylus* du sous-genre *Anomma* (Hymenoptera Formicidae). *Annales Musée Royal du Congo Belge. Nouvelle Série in Quarto, Sciences Zoologiques* 2: 1–359.

Raignier A, van Boven JKA, Ceusters R. 1974. Der Polymorphismus der afrikanischen Wanderameisen unter biometrischen und biologischen Gesichtspunkten. In *Sozialpolymorphismus bei Insekten. Probleme der Kastenbildung im Tierreich,* edited by GH Schmidt, 668–693. Stuttgart: Wissenschaftliche Verlagsgesellschaft.

Rakotonirina JC, Fisher BL. 2014. Revision of the Malagasy ponerine ants of the genus *Leptogenys* Roger (Hymenoptera: Formicidae). *Zootaxa* 3836: 1–163.

Ramírez S, Cameron SA. 2003. Army ant attacks by *Eciton hamatum* and *E. rapax* (Hymenoptera: Formicidae) on nests of the Amazonian bumble bee, *Bombus transversalis* (Hymenoptera: Apidae). *Journal of the Kansas Entomological Society* 76: 533–535.

Ramos-Elorduy de Concini J, Pino Moreno JM. 1988. The utilization of insects in the empirical medicine of ancient Mexicans. *Journal of Ethnobiology* 8: 195–202.

Rangel J, Mattila HR, Seeley TD. 2009. No intracolonial nepotism during colony fissioning in honey bees. *Proceedings of the Royal Society of London, Series B: Biological Sciences* 276: 3895–3900.

Ratnieks FLW, Wenseleers T. 2008. Altruism in insect societies and beyond: voluntary or enforced? *Trends in Ecology and Evolution* 23: 45–52.

Ratnieks FLW, Foster KR, Wenseleers T. 2006. Conflict resolution in insect societies. *Annual Review of Entomology* 51: 581–608.

Ravary F, Jaisson P. 2002. The reproductive cycle of thelytokous colonies of *Cerapachys biroi* Forel (Formicidae, Cerapachyinae). *Insectes Sociaux* 49: 114–119.

Ravary F, Jaisson P. 2004. Absence of individual sterility in thelytokous colonies of the ant *Cerapachys biroi* Forel (Formicidae, Cerapachyinae). *Insectes Sociaux* 51: 67–73.

Ravary F, Jahyny B, Jaisson P. 2006. Brood stimulation controls the phasic reproductive cycle of the parthenogenetic ant *Cerapachys biroi. Insectes Sociaux* 53: 20–26.

Ray TS, Andrews CC. 1980. Antbutterflies: butterflies that follow army ants to feed on antbird droppings. *Science* 210: 1147–1148.

Reeves DD, Moreau CS. 2019. The evolution of foraging behavior in ants (Hymenoptera: Formicidae). *Arthropod Systematics and Phylogeny* 77: 351–363.

Reichensperger A. 1923. Neue südamerikanische Histeriden als Gäste von Wanderameisen und Termiten. I. Systematischer Teil. *Mitteilungen der Schweizerischen Entomologischen Gesellschaft* 13: 313–336.

Reichensperger A. 1933. Ecitophilen aus Costa Rica (II), Brasilien und Peru (Staph. Hist. Clavig.). *Revista de Entomologia* 3: 179–194.

Reid CR, Lutz MJ, Powell S, Kao AB, Couzin ID, Garnier S. 2015. Army ants dynamically adjust living bridges in response to a cost-benefit trade-off. *Proceedings of the National Academy of Sciences of the United States of America* 112: 15113–15118.

Rettenmeyer CW. 1961. Observations on the biology and taxonomy of flies found over swarm raids of army ants (Diptera: Tachinidae, Conopidae). *University of Kansas Science Bulletin* 42: 993–1066.

Rettenmeyer CW. 1962a. The behavior of millipedes found with Neotropical army ants. *Journal of the Kansas Entomological Society* 35: 377–384.

Rettenmeyer CW. 1962b. Notes on host specificity and behavior of myrmecophilous macrochelid mites. *Journal of the Kansas Entomological Society* 35: 358–360.

Rettenmeyer CW. 1963a. Behavioral studies of army ants. *University of Kansas Science Bulletin* 44: 281–465.

Rettenmeyer CW. 1963b. The behavior of Thysanura found with army ants. *Annals of the Entomological Society of America* 56: 170–174.

Rettenmeyer CW. 1970. Insect mimicry. *Annual Review of Entomology* 15: 43–74.

Rettenmeyer CW, Akre RD. 1968. Ectosymbiosis between phorid flies and army ants. *Annals of the Entomological Society of America* 61: 1317–1326.

Rettenmeyer CW, Watkins JF II. 1978. Polygyny and monogyny in army ants (Hymenoptera: Formicidae). *Journal of the Kansas Entomological Society* 51: 581–591.

Rettenmeyer CW, Topoff H, Mirenda J. 1978. Queen retinues of army ants. *Annals of the Entomological Society of America* 71: 519–528.

Rettenmeyer CW, Chadab-Crepet R, Naumann MG, Morales L. 1983. Comparative foraging by Neotropical army ants. In *Social Insects in the Tropics,* Vol. 2. *Proceedings of the First International Symposium of the International Union for the Study of Social Insects and the Society for Mexican Entomology, Coyoyoc, Morelos, Mexico, November 1980,* edited by P Jaisson, 59–73. Paris: Université Paris-Nord.

Rettenmeyer CW, Rettenmeyer ME, Joseph J, Berghoff SM. 2011. The largest animal association centered on one species: the army ant *Eciton burchellii* and its more than 300 associates. *Insectes Sociaux* 58: 281–292.

Roberts DL, Cooper L, Petit J. 2000a. Flock characteristics of ant-following birds in premontane moist forest and coffee agroecosystems. *Ecological Applications* 10: 1414–1425.

Roberts DL, Cooper RJ, Petit LJ. 2000b. Use of premontane moist forest and shade coffee agroecosystems by army ants in western Panama. *Conservation Biology* 14: 192–199.

Rościszewski K, Maschwitz U. 1994. Prey specialization of army ants of the genus *Aenictus* in Malaysia. *Andrias* 13: 179–187.

Ryder Wilkie KT, Mertl AL, Traniello JFA. 2010. Species diversity and distribution patterns of the ants of Amazonian Ecuador. *PLoS One* 5: e13146.

Rylands AB, Monteiro da Cruz MAO, Ferrari SF. 1989. An association between marmosets and army ants in Brazil. *Journal of Tropical Ecology* 5: 113–116.

Sagili RR, Metz BN, Lucas HM, Chakrabarti P, Breece CR. 2018. Honey bees consider larval nutritional status rather than genetic relatedness when selecting larvae for emergency queen rearing. *Scientific Reports* 8: 7679.

Sánchez-Peña SR, Mueller UG. 2002. A nocturnal raid of *Nomamyrmex* army ants on *Atta* leaf-cutting ants in Tamaulipas, Mexico. *Southwestern Entomologist* 27: 221–223.

Santschi F. 1933. Contribution à l'étude des fourmis de l'Afrique tropicale. *Bulletin et Annales de la Société Entomologique de Belgique* 73: 95–108.

Sanz CM, Schöning C, Morgan DB. 2010. Chimpanzees prey on army ants with specialized tool set. *American Journal of Primatology* 72: 17–24.

Sasaki T, Pratt SC. 2018. The psychology of superorganisms: collective decision making by insect societies. *Annual Review of Entomology* 63: 259–275.

Satria R, Itioka T, Meleng P, Eguchi K. 2018. Second discovery of the subdichthadiigyne in *Yunodorylus* (Borowiec, 2009) (Formicidae: Dorylinae). *Revue Suisse de Zoologie* 125: 73–78.

Savage TS. 1847. On the habits of the "drivers" or visiting ants of West Africa. *Transactions of the Royal Entomological Society of London* 5: 1–15.

Sazima I. 2015. House Geckos (*Hemidactylus mabouia*) and an unidentified snake killed and devoured by army ants (*Eciton burchellii*). *Herpetology Notes* 8: 527–529.

Sazima I. 2017. New World army ants *Eciton burchellii* kill and consume leaf-litter inhabiting lizards in the Atlantic Forest, Southeast Brazil. *Tropical Natural History* 17: 119–122.

Schmid-Hempel P. 2011. *Evolutionary Parasitology: The Integrated Study of Infections, Immunology, Ecology, and Genetics.* Oxford: Oxford University Press.

Schmidt JO, Overal WL. 2009. Venom and task specialization in *Termitopone commutata* (Hymenoptera: Formicidae). *Journal of Hymenoptera Research* 18: 361–367.

Schmidt CA, Shattuck SO. 2014. The higher classification of the ant subfamily Ponerinae (Hymenoptera: Formicidae), with a review of ponerine ecology and behavior. *Zootaxa* 3817: 1–242.

Schmidt JO, Blum MS, Overal WL. 1986. Comparative enzymology of venoms from stinging Hymenoptera. *Toxicon* 24: 907–921.

Schneirla TC. 1933. Studies on army ants in Panama. *Journal of Comparative Psychology* 15: 267–299.

Schneirla TC. 1934. Raiding and other outstanding phenomena in the behavior of army ants. *Proceedings of the National Academy of Sciences of the United States of America* 20: 316–321.

Schneirla TC. 1938. A theory of army-ant behavior based upon the analysis of activities in a representative species. *Journal of Comparative Psychology* 25: 51–90.

Schneirla TC. 1940. Further studies on the army-ant behavior pattern: mass organization in the swarm-raiders. *Journal of Comparative Psychology* 29: 401–460.

Schneirla TC. 1944a. The reproductive functions of the army-ant queen as pace-makers of the group behavior pattern. *Journal of the New York Entomological Society* 52: 153–192.

Schneirla TC. 1944b. Studies on the army ant behavior pattern: nomadism in the swarm-raider *Eciton burchelli. Proceedings of the American Philosophical Society* 87: 438–457.

Schneirla TC. 1944c. A unique case of circular milling in ants, considered in relation to trail following and the general problem of orientation. *American Museum Novitates* 1253: 1–26.

Schneirla TC. 1945. The army-ant behavior pattern: nomad-statary relations in the swarmers and the problem of migration. *Biological Bulletin* 88: 166–193.

Schneirla TC. 1947. A study of army-ant life and behavior under dry-season conditions with special reference to reproductive functions. 1. Southern Mexico. *American Museum Novitates* 1336: 1–20.

Schneirla TC. 1948. Army-ant life and behavior under dry-season conditions with special reference to reproductive functions. 2. The appearance and fate of the males. *Zoologica* 33: 89–112.

Schneirla TC. 1949. Army-ant life and behavior under dry-season conditions. 3. The course of reproduction and colony behavior. *Bulletin of the American Museum of Natural History* 94: 7–81.

Schneirla TC. 1956. A preliminary survey of colony division and related processes in two species of terrestrial army ants. *Insectes Sociaux* 3: 49–69.

Schneirla TC. 1957a. A comparison of species and genera in the ant subfamily Dorylinae with respect to functional pattern. *Insectes Sociaux* 4: 259–298.

Schneirla TC. 1957b. Theoretical consideration of cyclic processes in doryline ants. *Proceedings of the American Philosophical Society* 101: 106–133.

Schneirla TC. 1958. The behavior and biology of certain Nearctic army ants: last part of the functional season, southeastern Arizona. *Insectes Sociaux* 5: 215–255.

Schneirla TC. 1961. The behavior and biology of certain Nearctic doryline ants. Sexual broods and colony division in *Neivamyrmex nigrescens. Zeitschrift für Tierpsychologie* 18: 1–32.

Schneirla TC. 1963. The behaviour and biology of certain Nearctic army ants: springtime resurgence of cyclic function—southeastern Arizona. *Animal Behaviour* 11: 583–595.

Schneirla TC. 1971. *Army Ants: A Study in Social Organization.* San Francisco: W. H. Freeman.

Schneirla TC, Brown RZ. 1950. Army-ant life and behavior under dry-season conditions. 4. Further investigations of cyclic processes in behavioral and reproductive functions. *Bulletin of the American Museum of Natural History* 95: 265–353.

Schneirla TC, Brown RZ. 1952. Sexual broods and the production of young queens in two species of army ants. *Zoologica* 37: 5–32.

Schneirla TC, Reyes AY. 1966. Raiding and related behaviour in two surface-adapted species of the Old World doryline ant *Aenictus. Animal Behaviour* 14: 132–148.

Schneirla TC, Reyes AY. 1969. Emigrations and related behaviour in two surface-adapted species of the Old World doryline ant *Aenictus. Animal Behaviour* 17: 87–103.

Schneirla TC, Brown RZ, Brown FC. 1954. The bivouac or temporary nest as an adaptive factor in certain terrestrial species of army ants. *Ecological Monographs* 24: 269–296.

Schneirla TC, Gianutsos RR, Pasternack BS. 1968. Comparative allometry of larval broods in three army-ant genera in relation to colony behavior. *American Naturalist* 102: 533–554.

Schöning C. 2008. Driver ants (*Dorylus* subgenus *Anomma*) (Hymenoptera: Formicidae). In *Encyclopedia of Entomology,* edited by JL Capinera. Dordrecht, the Netherlands: Springer.

Schöning C, Moffett MW. 2007. Driver ants invading a termite nest: why do the most catholic predators of all seldom take this abundant prey? *Biotropica* 39: 663–667.

Schöning C, Kinuthia W, Franks NR. 2005a. Evolution of allometries in the worker caste of *Dorylus* army ants. *Oikos* 110: 231–240.

Schöning C, Njagi WM, Franks NR. 2005b. Temporal and spatial patterns in the emigrations of the army ant *Dorylus* (*Anomma*) *molestus* in the montane forest of Mt Kenya. *Ecological Entomology* 30: 532–540.

Schöning C, Kinuthia W, Boomsma JJ. 2006. Does the afrotropical army ant *Dorylus* (*Anomma*) *molestus* go extinct in fragmented forests? *Journal of East African Natural History* 95: 163–179.

Schöning C, Ellis D, Fowler A, Sommer V. 2007. Army ant prey availability and consumption by chimpanzees (*Pan troglodytes vellerosus*) at Gashaka / Nigeria. *Journal of Zoology* 271: 125–133.

Schöning C, Gotwald WH Jr, Kronauer DJC, Vilhelmsen L. 2008a. Taxonomy of the African army ant *Dorylus gribodoi* Emery, 1892 (Hymenoptera, Formicidae)—new insights from DNA sequence data and morphology. *Zootaxa* 1749: 39–52.

Schöning C, Humle T, Möbius Y, McGrew WC. 2008b. The nature of culture: technological variation in chimpanzee predation on army ants revisited. *Journal of Human Evolution* 55: 48–59.

Schöning C, Njagi W, Kinuthia W. 2008c. Prey spectra of two swarm-raiding army ant species in East Africa. *Journal of Zoology* 274: 85–93.

Schöning C, Csuzdi C, Kinuthia W. 2010. Influence of driver ant swarm raids on earthworm prey densities in the Mount Kenya forest: implications for prey population dynamics and colony migrations. *Insectes Sociaux* 57: 73–82.

Schöning C, Shepard L, Sen A, Kinuthia W, Ogutu JO. 2011. Should I stay or should I go now? Patch use by African army ant colonies in relation to food availability and predation. *Insectes Sociaux* 58: 335–343.

Seeley TD. 2010. *Honeybee Democracy.* Princeton, NJ: Princeton University Press.

Seevers CH. 1965. The systematics, evolution and zoogeography of staphylinid beetles associated with army ants (Coleoptera, Staphylinidae). *Fieldiana Zoology* 47: 137–351.

Seignobos C, Deguine J-P, Aberlenc H-P. 1996. Les Mofus et leurs insectes. *Journal d'Agriculture Traditionnelle et de Botanique Appliquée* 38: 125–187.

Shattuck SO. 2008. Review of the ant genus *Aenictus* (Hymenoptera: Formicidae) in Australia with notes on *A. ceylonicus* (Mayr). *Zootaxa* 1926: 1–19.

Slessor KN, Winston ML, Le Conte Y. 2005. Pheromone communication in the honeybee (*Apis mellifera* L.). *Journal of Chemical Ecology* 31: 2731–2745.

Smith F. 1855. XVII. Descriptions of some species of Brazilian ants belonging to the genera *Pseudomyrma, Eciton* and *Myrmica* (with observations on their economy by Mr. H. W. Bates). *Transactions of the Royal Entomological Society of London* 8: 156–169.

Smith F. 1858. *Catalogue of Hymenopterous Insects in the Collection of the British Museum. Part VI. Formicidae.* London: Taylor and Francis.

Smith F. 1860. VI. Descriptions of new genera and species of exotic Hymenoptera. *Journal of Entomology* 1: 65–84.

Smith MR. 1942. The legionary ants of the United States belonging to *Eciton* subgenus *Neivamyrmex* Borgmeier. *American Midland Naturalist* 27: 537–590.

Smith AA, Haight KL. 2008. Army ants as research and collection tools. *Journal of Insect Science* 8: 71.

Snelling GC, Snelling RR. 2007. New synonymy, new species, new keys to *Neivamyrmex* army ants of the United States. In *Advances in Ant Systematics (Hymenoptera: Formicidae): Homage to E. O. Wilson: 50 Years of Contributions,* edited by RR Snelling, BL Fisher, PS Ward, 459–550. Memoirs of the American Entomological Institute, vol. 80. Gainesville, FL: American Entomological Institute.

Soare TW, Tully SI, Willson SK, Kronauer DJC, O'Donnell SO. 2011. Choice of nest site protects army ant colonies from environmental extremes in tropical montane forest. *Insectes Sociaux* 58: 299–308.

Soare TW, Kumar A, Naish KA, O'Donnell S. 2014. Genetic evidence for landscape effects on dispersal in the army ant *Eciton burchellii. Molecular Ecology* 23: 96–109.

Soare TW, Kumar A, Naish KA, O'Donnell S. 2020. Multi-year genetic sampling indicates maternal gene flow via colony emigrations in the army ant *Eciton burchellii parvispinum. Insectes Sociaux* 67: 155–166.

Solé RV, Bonabeau E, Delgado J, Fernandez P, Martin J. 2000. Pattern formation and optimisation in army ant raids. *Artificial Life* 6: 219–227.

Souza JLP, Moura CAR. 2008. Predation of ants and termites by army ants, *Nomamyrmex esenbeckii* (Formicidae, Ecitoninae) in the Brazilian Amazon. *Sociobiology* 52: 399–402.

Sprenger PP, Menzel F. 2020. Cuticular hydrocarbons in ants (Hymenoptera: Formicidae) and other insects: how and why they differ among individuals, colonies, and species. *Myrmecological News* 30: 1–26.

Staab M. 2014. The first observation of honeydew foraging in army ants since 1933: *Aenictus hodgsoni* Forel, 1901 tending *Eutrichosiphum heterotrichum* (Raychaudhuri, 1956) in Southeast China. *Asian Myrmecology* 6: 115–118.

Step E. 1916. *Marvels of Insect Life.* New York: Robert M. McBride & Company.

Stephenson C. 1938. Leiningen versus the ants. *Esquire* 10: 98–99, 235–241.

Stouffer PC, Bierregaard RO Jr. 1995. Use of Amazonian forest fragments by understory insectivorous birds. *Ecology* 76: 2429–2445.

Stouffer PC, Bierregaard RO Jr, Strong C, Lovejoy TE. 2006. Long-term landscape change and bird abundance in Amazonian rainforest fragments. *Conservation Biology* 20: 1212–1223.

Sudd JH. 1972. Reviewed Work: Army ants, a study in social organization. *Science Progress* 60: 570–573.

Sugiyama Y. 1995. Tool-use for catching ants by chimpanzees at Bossou and Monts Nimba, West Africa. *Primates* 36: 193–205.

Swartz MB. 1997. Behavioral and population ecology of the army ant *Eciton burchelli* and ant-following birds. Ph.D. diss., University of Texas, Austin.

Swartz MB. 1998. Predation on an *Atta cephalotes* colony by an army ant, *Nomamyrmex esenbeckii. Biotropica* 30: 682–684.

Swartz MB. 2001. Bivouac checking, a novel behavior distinguishing obligate from opportunistic species of army-ant-following birds. *The Condor* 103: 629–633.

Talbot M, Kennedy CH. 1940. The slave-making ant, *Formica sanguinea subintegra* Emery, its raids, nuptial flights and nest structure. *Annals of the Entomological Society of America* 33: 560–577.

Tafuri JF. 1955. Growth and polymorphism in the larva of the army ant (*Eciton (E.) hamatum* Fabricius). *Journal of the New York Entomological Society* 63: 21–41.

Tarpy DR, Gilley DC, Seeley TD. 2004. Levels of selection in a social insect: a review of conflict and cooperation during honey bee (*Apis mellifera*) queen replacement. *Behavioral Ecology and Sociobiology* 55: 513–523.

Teles da Silva M. 1977a. Behaviour of the army ant *Eciton burchelli* Westwood (Hymenoptera: Formicidae) in the Belém region. I. Nomadic-statary cycles. *Animal Behaviour* 25: 910–923.

Teles da Silva M. 1977b. Behavior of the army ant *Eciton burchelli* Westwood (Hymenoptera Formicidae) in the Belém region. II. Bivouacs. *Boletim de Zoologia, Universidade de São Paulo* 2: 107–128.

Teles da Silva M. 1982. Behaviour of army ants *Eciton burchellii* and *E. hamatum* (Hymenoptera, Formicidae) in the Belem region III. Raid activity. *Insectes Sociaux* 29: 243–267.

Teseo S, Delloro F. 2017. Reduced foraging investment as an adaptation to patchy food sources: a phasic army ant simulation. *Journal of Theoretical Biology* 428: 48–55.

Teseo S, Kronauer DJC, Jaisson P, Châline N. 2013. Enforcement of reproductive synchrony via policing in a clonal ant. *Current Biology* 23: 328–332.

Thomas JA, Schönrogge K, Elmes GW. 2005. Specializations and host associations of social parasites of ants. In *Insect Evolutionary Ecology,* edited by MDE Fellowes, GJ Holloway, J Rolff, 479–518. Wallingford, United Kingdom: CAB International.

Tishechkin AK, Kronauer DJC, von Beeren C. 2017. Taxonomic review and natural history notes of the army ant-associated beetle genus *Ecclisister* Reichensperger (Coleoptera: Histeridae: Haeteriinae). *Coleopterists Bulletin* 71: 279–288.

Topoff H. 1971. Polymorphism in army ants related to division of labor and colony cyclic behavior. *American Naturalist* 105: 529–548.

Topoff H. 1975. Behavioral changes in the army ant *Neivamyrmex nigrescens* during the nomadic and statary phases. *Journal of the New York Entomological Society* 83: 38–48.

Topoff H. 1984. Social organization of raiding and emigrations in army ants. *Advances in the Study of Behavior* 14: 81–126.

Topoff H, Lawson K. 1979. Orientation of the army ant *Neivamyrmex nigrescens:* integration of chemical and tactile information. *Animal Behaviour* 27: 429–433.

Topoff H, Mirenda J. 1975. Trail-following by the army ant *Neivamyrmex nigrescens:* responses by workers to volatile odors. *Annals of the Entomological Society of America* 68: 1044–1046.

Topoff H, Mirenda J. 1978. Precocial behaviour of callow workers of the army ant *Neivamyrmex nigrescens:* importance of stimulation by adults during mass recruitment. *Animal Behaviour* 26: 698–706.

Topoff H, Mirenda J. 1980a. Army ants do not eat and run: influence of food supply on emigration behaviour in *Neivamyrmex nigrescens. Animal Behaviour* 28: 1040–1045.

Topoff H, Mirenda J. 1980b. Army ants on the move: the relationship between food supply and migration frequency. *Science* 207: 1099–1100.

Topoff H, Boshes M, Trakimas W. 1972a. A comparison of trail following between callow and adult workers of the army ant [*Neivamyrmex nigrescens* (Formicidae: Dorylinae)]. *Animal Behaviour* 20: 361–366.

Topoff H, Lawson K, Richards P. 1972b. Trail following and its development in the Neotropical army ant genus *Eciton* (Hymenoptera: Formicidae: Dorylinae). *Psyche* 79: 357–364.

Topoff H, Lawson K, Richards P. 1973. Trail following in two species of the army ant genus *Eciton:* comparison between major and intermediate-sized workers. *Annals of the Entomological Society of America* 66: 109–111.

Topoff H, Mirenda J, Droual R, Herrick S. 1980a. Behavioural ecology of mass recruitment in the army ant *Neivamyrmex nigrescens. Animal Behaviour* 28: 779–786.

Topoff H, Mirenda J, Droual R, Herrick S. 1980b. Onset of the nomadic phase in the army ant *Neivamyrmex nigrescens* (Cresson) (Hym. Form.): distinguishing between callow and larval excitation by brood substitution. *Insectes Sociaux* 27: 175–179.

Topoff H, Rothstein A, Pujdak S, Dahlstrom T. 1981. Statary behavior in nomadic colonies of army ants: the effect of overfeeding. *Psyche* 88: 151–161.

Torgerson RL, Akre RD. 1969. Reproductive morphology and behavior of a Thysanuran, *Trichatelura manni,* associated with army ants. *Annals of the Entomological Society of America* 62: 1367–1374.

Torgerson RL, Akre RD. 1970. The persistence of army ant chemical trails and their significance in the Ecitonine-Ecitophile association (Formicidae: Ecitonini). *Melanderia* 5: 1–28.

Tórrez MA, Arendt W, Salmeron P. 2009. Aves hormigueras en bosque seco del Pacífico de Nicaragua: uso de hábitat y comportamiento parasítico. *Zeledonia* 13: 1–9.

Trible W, Kronauer DJC. 2017. Caste development and evolution in ants: it's all about size. *Journal of Experimental Biology* 220: 53–62.

Trible W, Olivos-Cisneros L, McKenzie SK, Saragosti J, Chang N-C, Matthews BJ, Oxley PR, Kronauer DJC. 2017. *orco* mutagenesis causes loss of antennal lobe glomeruli and impaired social behavior in ants. *Cell* 170: 727–735.

Tsuji K, Yamauchi K. 1995. Production of females by parthenogenesis in the ant, *Cerapachys biroi. Insectes Sociaux* 42: 333–336.

Tsutsui ND, Suarez AV. 2003. The colony structure and population biology of invasive ants. *Conservation Biology* 17: 48–58.

Ulrich Y, Burns D, Libbrecht R, Kronauer DJC. 2016. Ant larvae regulate worker foraging behavior and ovarian activity in a dose-dependent manner. *Behavioral Ecology and Sociobiology* 70: 1011–1018.

Ulrich Y, Saragosti J, Tokita CK, Tarnita CE, Kronauer DJC. 2018. Fitness benefits and emergent division of labor at the onset of group-living. *Nature* 560: 635–638.

Ulrich Y, Kawakatsu M, Tokita CK, Saragosti J, Chandra V, Tarnita CE, Kronauer DJC. 2020. Emergent behavioral organization in heterogeneous groups of a social insect. *bioRxiv* 2020.03.05.963207. Available at https//doi.org/10.1101/2020.03.05.963207.

Vallely AC. 2001. Foraging at army ant swarms by fifty bird species in the highlands of Costa Rica. *Ornitología Neotropical* 12: 271–275.

van Wilgenburg E, Driessen G, Beukeboom LW. 2006. Single locus complementary sex determination in Hymenoptera: an "unintelligent" design? *Frontiers in Zoology* 3: 1.

van Zweden JS, d'Ettorre P. 2010. Nestmate recognition in social insects and the role of hydrocarbons. In *Insect Hydrocarbons: Biology, Biochemistry, and Chemical Ecology,* edited by GJ Blomquist, AG Bagnères, 222–243. Cambridge: Cambridge University Press.

Varela-Hernández F, Castaño-Meneses G. 2011. *Neivamyrmex albacorpus*, a new species of ant (Hymenoptera: Formicidae: Ecitoninae) from Metztitlán, Hidalgo, México. *Sociobiology* 58: 579–584.

Vidal-Riggs JM, Chaves-Campos J. 2008. Method review: estimation of colony densities of the army ant *Eciton burchellii* in Costa Rica. *Biotropica* 40: 259–262.

Vieira RS, Höfer H. 1994. Prey spectrum of two army ant species in central Amazonia, with special attention on their effect on spider populations. *Andrias* 13: 189–198.

Villet MH. 1990. Division of labour in the Matabele ant *Megaponera foetens* (Fabr.) (Hymenoptera Formicidae). *Ethology Ecology and Evolution* 2: 397–417.

Visscher PK. 2007. Group decision making in nest-site selection among social insects. *Annual Review of Entomology* 52: 255–275.

von Beeren C, Tishechkin AK. 2017. *Nymphister kronaueri* von Beeren & Tishechkin sp. nov., an army ant-associated beetle species (Coleoptera: Histeridae: Haeteriinae) with an exceptional mechanism of phoresy. *BMC Zoology* 2: 3.

von Beeren C, Maruyama M, Hashim R, Witte V. 2011a. Differential host defense against multiple parasites in ants. *Evolutionary Ecology* 25: 259–276.

von Beeren C, Schulz S, Hashim R, Witte V. 2011b. Acquisition of chemical recognition cues facilitates integration into ant societies. *BMC Ecology* 11: 30.

von Beeren C, Hashim R, Witte V. 2012a. The social integration of a myrmecophilous spider does not depend exclusively on chemical mimicry. *Journal of Chemical Ecology* 38: 262–271.

von Beeren C, Pohl S, Witte V. 2012b. On the use of adaptive resemblance terms in chemical ecology. *Psyche* 2012: 635761.

von Beeren C, á l'Allemand SL, Hashim R, Witte V. 2014. Collective exploitation of a temporally unpredictable food source: mushroom harvesting by the ant *Euprenolepis procera*. *Animal Behaviour* 89: 71–77.

von Beeren C, Maruyama M, Kronauer DJC. 2016a. Community sampling and integrative taxonomy reveal new species and host specificity in the army ant-associated beetle genus *Tetradonia* (Coleoptera, Staphylinidae, Aleocharinae). *PLoS One* 11: e0165056.

von Beeren C, Maruyama M, Kronauer DJC. 2016b. Cryptic diversity, high host specificity and reproductive synchronization in army ant-associated *Vatesus* beetles. *Molecular Ecology* 25: 990–1005.

von Beeren C, Brückner A, Maruyama M, Burke G, Wieschollek J, Kronauer DJC. 2018. Chemical and behavioral integration of army ant-associated rove beetles—a comparison between specialists and generalists. *Frontiers in Zoology* 15: 8.

von Beeren C, Blüthgen N, Brückner A, Hoenle PO, Tishechkin AK, Maruyama M, Brown B, Hash J, Hall WE, Pohl S, Ospina Jara B, Kronauer DJC. n.d. Symbiont communities of army ants: host specificity and its relation to behavioral and chemical integration. Unpublished manuscript.

Wallace AR. 1878. *Tropical Nature, and Other Essays*. London: McMillan and Co.

Wang YJ, Happ GM. 1974. Larval development during the nomadic phase of a Nearctic army ant, *Neivamyrmex nigrescens* (Cresson) (Hymenoptera: Formicidae). *International Journal of Insect Morphology and Embryology* 3: 73–86.

Ward PS. 2007. The ant genus *Leptanilloides:* discovery of the male and evaluation of phylogenetic relationships based on DNA sequence data. In *Advances in Ant Systematics (Hymenoptera: Formicidae): Homage to E. O. Wilson: 50 Years of Contributions,* edited by RR Snelling, BL Fisher, PS Ward, 637–649. Memoirs of the American Entomological Institute, vol. 80. Gainesville, FL: American Entomological Institute.

Ward PS. 2014. The phylogeny and evolution of ants. *Annual Review of Ecology, Evolution, and Systematics* 45: 23–43.

Wasmann E. 1887. Neue brasilianische Staphyliniden, bei *Eciton hamatum* gesammelt von Dr. W. Müller. *Deutsche Entomologische Zeitschrift* 31: 403–416.

Wasmann E. 1895. Die Ameisen- und Termitengäste von Brasilien. *Verhandlungen der kaiserlich-königlichen zoologisch-botanischen Gesellschaft in Wien* 45: 137–178.

Wasmann E. 1903. Zur näheren Kenntnis des echten Gastverhältnisses (Symphilie) bei den Ameisen und Termitengästen. *Biologisches Zentralblatt* 23: 63–72, 195–207, 232–248, 261–276, 298–310.

Wasmann E. 1904. Zur Kenntnis der Gäste der Treiberameisen und ihrer Wirthe am oberen Congo, nach den Sammlungen und Beobachtungen von P. Herm. Kohl, C.S.S.C. bearbeitet. Supplement. *Zoologische Jahrbücher* 7: 611–682.

Wasmann E. 1925. *Die Ameisenmimikry: Ein exakter Beitrag zum Mimikryproblem und zur Theorie der Anpassung*. Abhandlungen zur theoretischen Biologie 19. Berlin: Gebrüder Bornträger.

Watkins JF II. 1964. Laboratory experiments on the trail following of army ants of the genus *Neivamyrmex* (Formicidae: Dorylinae). *Journal of the Kansas Entomological Society* 37: 22–28.

Watkins JF II. 1976. *The Identification and Distribution of New World Army Ants (Dorylinae: Formicidae)*. Waco, TX: Baylor University Press.

Watkins JF II. 1985. The identification and distribution of the army ants of the United States of America (Hymenoptera, Formicidae, Ecitoninae). *Journal of the Kansas Entomological Society* 58: 479–502.

Watkins JF II, Cole TW. 1966. The attraction of army ant workers to secretions of their queen. *Texas Journal of Science* 18: 254–265.

Watkins JF II, Cole TW, Baldridge RS. 1967a. Laboratory studies on interspecific trail following and trail preference of army ants (Dorylinae). *Journal of the Kansas Entomological Society* 40: 146–151.

Watkins JF II, Gehlbach FR, Baldridge RS. 1967b. Ability of the blind snake, *Leptotyphlops dulcis,* to follow pheromone trails of army ants, *Neivamyrmex nigrescens* and *N. opacithorax. Southwestern Naturalist* 12: 455–462.

Watkins JF II, Gehlbach FR, Kroll JC. 1969. Attractant-repellant secretions in blind snakes (*Leptotyphlops dulcis*) and army ants (*Neivamyrmex nigrescens*). *Ecology* 50: 1098–1102.

Watkins JF II, Gehlbach FR, Plsek RW. 1972. Behavior of blind snakes (*Leptotyphlops dulcis*) in response to army ant (*Neivamyrmex nigrescens*) raiding columns. *Texas Journal of Science* 23: 556–557.

Weber NA. 1941. The rediscovery of the queen of *Eciton (Labidus) coecum* Latr. *American Midland Naturalist* 26: 325–329.

Weissflog A, Maschwitz U, Disney RHL, Rościszewski K. 1995. A fly's ultimate con. *Nature* 378: 137.

Weissflog A, Sternheim E, Dorow WHO, Berghoff S, Maschwitz U. 2000. How to study subterranean army ants: a novel method for locating and monitoring field populations of the South East Asian army ant *Dorylus (Dichthadia) laevigatus* Smith, 1857 (Formicidae, Dorylinae) with observations on their ecology. *Insectes Sociaux* 47: 317–324.

Werringloer A. 1932. Die Sehorgane und Sehzentren der Dorylinen nebst Untersuchungen über die Facettenaugen der Formiciden. *Zeitschrift für wissenschaftliche Zoologie* 141: 432–524.

West-Eberhard MJ. 1989. Scent-trail diversion, a novel defense against ants by tropical social wasps. *Biotropica* 21: 280–281.

Westwood JO. 1842. Monograph of the hymenopterous group, Dorylides. In *Arcana Entomologica; or Illustrations of New, Rare, and Interesting Insects,* 73–80. Vol. 1, No. 5. London: W. Smith.

Wheeler WM. 1903. Some notes on the habits of *Cerapachys augustae. Psyche* 10: 205–209.

Wheeler WM. 1908. The polymorphism of ants. *Annals of the Entomological Society of America* 1: 39–69.

Wheeler WM. 1910. *Ants—Their Structure, Development, and Behavior.* New York: Columbia University Press.

Wheeler WM. 1911. The ant-colony as an organism. *Journal of Morphology* 22: 307–325.

Wheeler WM. 1918. The Australian ants of the ponerine tribe Cerapachyini. *Proceedings of the American Academy of Arts and Sciences* 53: 215–265.

Wheeler WM. 1921. Observations on army ants in British Guiana. *Proceedings of the American Academy of Arts and Sciences* 56: 291–328.

Wheeler WM. 1925. The finding of the queen of the army ant *Eciton hamatum* Fabricius. *Biological Bulletin* 49: 139–149.

Wheeler GC. 1943. The larvae of the army ants. *Annals of the Entomological Society of America* 36: 319–332.

Wheeler GC, Wheeler J. 1964. The ant larvae of the subfamily Dorylinae: supplement. *Proceedings of the Entomological Society of Washington* 66: 129–137.

Wheeler GC, Wheeler J. 1974. Ant larvae of the subfamily Dorylinae: second supplement. *Journal of the Kansas Entomological Society* 47: 166–172.

Wheeler GC, Wheeler J. 1984. The larvae of the army ants (Hymenoptera: Formicidae): a revision. *Journal of the Kansas Entomological Society* 57: 263–275.

Wheeler GC, Wheeler J. 1986. Young larvae of *Eciton* (Hymenoptera: Formicidae: Dorylinae). *Psyche* 93: 341–349.

Whelden RM. 1963. The anatomy of the adult queen and workers of the army ants *Eciton burchelli* Westwood and *Eciton hamatum* Fabricus. *Journal of the New York Entomological Society* 71: 158–178.

Wiley RH. 1980. Multispecies antbird societies in lowland forest of Surinam and Ecuador: stable membership and foraging differences. *Journal of the Zoological Society of London* 191:127–145.

Willis EO. 1967. The behavior of bicolored antbirds. *University of California Publications in Zoology* 79: 1–132.

Willis EO. 1973. The behavior of ocellated antbirds. *Smithsonian Contributions to Zoology* 144: 1–57.

Willis EO, Oniki Y. 1978. Birds and army ants. *Annual Review of Ecology and Systematics* 9: 243–263.

Willson SK. 2004. Obligate army-ant-following birds: a study of ecology, spatial movement patterns, and behavior in Amazonian Peru. *Ornithological Monographs* 55: 1–67.

Willson SK, Sharp R, Ramler IP, Sen A. 2011. Spatial movement optimization in Amazonian *Eciton burchellii* army ants. *Insectes Sociaux* 58: 325–334.

Wilson EO. 1958a. The beginnings of nomadic and group-predatory behavior in the ponerine ants. *Evolution* 12: 24–36.

Wilson EO. 1958b. Observations on the behavior of the cerapachyine ants. *Insectes Sociaux* 5: 129–140.

Wilson EO. 1959. Studies on the ant fauna of Melanesia VI. The tribe Cerapachyini. *Pacific Insects* 1: 39–57.

Wilson EO. 1964. The true army ants of the Indo-Australian area (Hymenoptera: Formicidae: Dorylinae). *Pacific Insects* 6: 427–483.

Wilson EO. 1971. *The Insect Societies.* Cambridge, MA: Belknap Press.

Wilson EO. 1975. *Sociobiology: The New Synthesis.* Cambridge, MA: Harvard University Press.

Wilson EO. 1985. Ants of the Dominican amber (Hymenoptera: Formicidae). 2. The first fossil army ants. *Psyche* 92: 11–16.

Wilson EO. 1990. *Success and Dominance in Ecosystems: The Case of the Social Insects.* Oldendorf / Luhe: Ecology Institute.

Wilson EO, Hölldobler B. 2005. The rise of the ants: a phylogenetic and ecological explanation. *Proceedings of the National Academy of Sciences of the United States of America* 102: 7411–7414.

Wilson EO, Eisner T, Valentine BD. 1954. The beetle genus *Paralimulodes* Bruch in North America, with notes on morphology and behavior (Coleoptera: Limulodidae). *Psyche* 61: 154–161.

Wilson EO, Gómez Durán JM. 2010. *Kingdom of Ants: José Celestino Mutis and the Dawn of Natural History in the New World.* Baltimore: Johns Hopkins University Press.

Winston ML. 1991. *The Biology of the Honey Bee.* Cambridge, MA: Harvard University Press.

Winston ME, Kronauer DJC, Moreau CS. 2017. Early and dynamic colonization of Central America drives speciation in Neotropical army ants. *Molecular Ecology* 26: 859–870.

Witte V, Maschwitz U. 2000. Raiding and emigration dynamics in the ponerine army ant *Leptogenys distinguenda* (Hymenoptera, Formicidae). *Insectes Sociaux* 47: 76–83.

Witte V, Maschwitz U. 2002. Coordination of raiding and emigration in the ponerine army ant *Leptogenys distinguenda* (Hymenoptera: Formicidae: Ponerinae): a signal analysis. *Journal of Insect Behavior* 15: 195–217.

Witte V, Maschwitz U. 2008. Mushroom harvesting ants in the tropical rain forest. *Naturwissenschaften* 95: 1477–1486.

Witte V, Hänel H, Weissflog A, Rosli H, Maschwitz U. 1999. Social integration of the myrmecophilic spider *Gamasomorpha maschwitzi* (Araneae: Oonopidae) in colonies of the South East Asian army ant, *Leptogenys distinguenda* (Formicidae: Ponerinae). *Sociobiology* 34: 145–159.

Witte V, Janssen R, Eppenstein A, Maschwitz U. 2002. *Allopeas myrmekophilos* (Gastropoda, Pulmonata), the first myrmecophilous mollusc living in colonies of the ponerine army ant *Leptogenys distinguenda* (Formicidae, Ponerinae). *Insectes Sociaux* 49: 301–305.

Witte V, Leingärtner A, Sabaß L, Hashim R, Foitzik S. 2008. Symbiont microcosm in an ant society and the diversity of interspecific interactions. *Animal Behaviour* 76: 1477–1486.

Witte V, Foitzik S, Hashim R, Maschwitz U, Schulz S. 2009. Fine tuning of social integration by two myrmecophiles of the ponerine army ant, *Leptogenys distinguenda*. *Journal of Chemical Ecology* 35: 355–367.

Witte V, Schliessmann D, Hashim R. 2010. Attack or call for help? Rapid individual decisions in a group-hunting ant. *Behavioral Ecology* 21: 1040–1047.

Wrege PH, Wikelski M, Mandel JT, Rassweiler T, Couzin ID. 2005. Antbirds parasitize foraging army ants. *Ecology* 86: 555–559.

Wunderlich J. 1994. Beschreibung bisher unbekannter Spinnenarten und Gattungen aus Malaysia und Indonesien (Arachnida: Araneae: Oonopidae, Tetrablemidae, Telemidae, Pholcidae, Linyphiidae, Nesticidae, Theridiidae und Dictynidae). *Beiträge zur Araneologie* 4: 559–580.

Xu ZH. 2000. Two new genera of ant subfamilies Dorylinae and Ponerinae (Hymenoptera: Formicidae) from Yunnan, China. *Zoological Research* 21: 297–302.

Yamane S, Hashimoto Y. 1999. A remarkable new species of the army ant genus *Aenictus* (Hymenoptera, Formicidae) with a polymorphic worker caste. *Tropics* 8: 427–432.

Young AM. 1979. Attacks by the army ant *Eciton burchellii* on nests of the social paper wasp *Polistes erythrocephalus* in northeastern Costa Rica. *Journal of the Kansas Entomological Society* 52: 759–768.

Yusuf AA, Crewe RM, Pirk CWW. 2014a. Olfactory detection of prey by the termite-raiding ant *Pachycondyla analis*. *Journal of Insect Science* 14: 53.

Yusuf AA, Gordon I, Crewe RM, Pirk CWW. 2014b. Prey choice and raiding behaviour of the ponerine ant *Pachycondyla analis* (Hymenoptera: Formicidae). *Journal of Natural History* 48: 345–358.

Zhou Y-L, Ślipiński A, Ren D, Parker J. 2019. A Mesozoic clown beetle myrmecophile (Coleoptera: Histeridae). *eLife* 8: e44985.

acknowledgments

The idea to write a book about army ants goes back more than a decade, and in retrospect I realize that the history of these fabulous creatures has intersected with my own on numerous occasions throughout my life. I was born in Heidelberg, a quaint German college town with a rich and ancient history. Among the most famous depictions of Heidelberg with its castle and Old Bridge is a copper engraving by Matthäus Merian from 1620. Merian's daughter, Maria Sibylla Merian, was fascinated with insect metamorphosis as a child and later became one of the most significant entomologists and naturalists of her time. She would provide what is arguably the first detailed description of army ants based on observations she made in Suriname. My own fascination with the living world also became apparent early on. As a kid, I acquired numerous crawling roommates from the local woods and the exotic pet trade while immersing myself in books about faraway tropical rainforests, dreaming of my own expeditions that surely were to come. I am most grateful for my wonderful parents, Usch and Ulrich Kronauer, who always let me explore freely and whose unconditional love has given me the confidence to take risks in my personal and professional life. I also thank my sisters, Lisa and Marie, for their companionship as children in Heidelberg and now as adults scattered across the globe.

Having completed my first two years of study at the University of Heidelberg, I moved to the University of Würzburg, initially with the intent to study tropical ecology. However, I was soon drawn to the Department of Behavioral Physiology and Sociobiology, headed by the famous ant expert Bert Hölldobler and staffed with a suite of group leaders working on different aspects of ant and bee biology. Among them was Jürgen Gadau, a vibrant young professor who was capitalizing on a recent revolution in biology. By the turn of the millennium, DNA sequencing and genotyping had become widely available to study essentially any organism one was interested in, and Jürgen was applying these techniques to study the ecology and evolution of ants in ways that, until quite recently, had been unthinkable. I had found my calling, and once I had started to work with Bert and Jürgen, initially on the evolution of slave-making behavior in honeypot ants, I never looked back. I thank Bert and Jürgen for their inspiring work and mentorship, and for passing on their love of ants to me. The time I spent with them in Würzburg certainly had a tremendous impact on my life's trajectory.

It was also in Würzburg where I met Koos Boomsma, a professor at the University of Copenhagen and renowned specialist on social evolution in general and leaf-cutting ants in particular, who was there on sabbatical. I soon approached Koos about the possibility of pursuing graduate studies in his laboratory, assuming that I might work on leaf-cutting ants as well. But Koos had a better idea. Knowing that I wanted to do a lot of fieldwork in the tropics, he suggested I turn to army ants instead. That evening I went home and read the chapter on army ants in the tome *The Ants* by Bert Hölldobler and Edward O. Wilson. I was mesmerized. The next day, back at work, I began to gather all the relevant literature I could find and kept reading. It dawned on me, and I am sure Koos had already been aware of this, that nobody had ever applied genetic approaches to the study of army ants, and that the opportunities were thus enormous. The project for my PhD thesis was born.

In December 2003, I moved to Copenhagen to join Koos's group, and over the following three years I traveled to Venezuela, Kenya, Costa Rica, and Kansas to observe and collect army ants, followed by days and nights in the laboratory extracting army ant DNA and genotyping hundreds and hundreds of specimens. I was doing exactly what I had dreamed of as a child back in Heidelberg, but even better than that: I was able to follow up on my natural history expeditions with modern genetic tools. To my great surprise, it eventually turned out that the University of Copenhagen had been intricately linked to the study of army ants long before my arrival. During a visit to the neighboring Zoological Museum with Edith Rodríguez, an ant taxonomist at the University of Venezuela and my field companion for several years, we happened across a crucial army ant type specimen (see Chapter 1 for the full story). This specimen had been described by the Danish zoologist Johan Christian Fabricius in 1782 but had been believed to be missing for more than half a century.

Toward the end of my doctoral work, Koos approached me with a bold proposition: he offered to let me stay for an extra year to write a book about army ants. Loving the idea, I agreed to stay in Copenhagen, but I soon realized that I wasn't quite ready to write the kind of book I would have wanted. Instead, I dedicated the extra time to more laboratory work, to finishing research projects and starting new ones. I thank Koos for setting me loose on army ants, for creating a uniquely nurturing research environment in Copenhagen, for his generosity in sharing ideas and allowing me to pursue my own freely, and for planting the idea to write a book about army ants in my head.

In 2008, I moved to the University of Lausanne for a brief postdoctoral stint in the laboratory of Laurent Keller, another famous myrmecologist. Laurent is arguably one of the clearest thinkers in our field, and our friendship and many critical discussions have been immensely valuable and inspiring. While working in Lausanne, I lived in Morges, just a couple of kilometers from campus and the birthplace of the Swiss psychiatrist and myrmecologist Auguste Forel, whose name lives on in one of the main protagonists of this book, the army ant *Eciton burchellii foreli.* This is the subspecies of *Eciton burchellii* present at La Selva Biological Station in Costa Rica.

After a wonderful summer in Lausanne, I moved again, this time across the Atlantic to the Museum of Comparative Zoology at Harvard University. The museum's ant collection is somewhat of a sacred site for ant enthusiasts, breathing history and teeming with precious specimens. The collection was founded in 1908 by William Morton Wheeler, and it was Wheeler who discovered and described the queens of *Eciton burchellii* and *Eciton hamatum,* my second main protagonist. It is also here where Edward O. Wilson developed a first outline for the evolution of army ant behavior. During my time at Harvard, I worked in the laboratory of Naomi Pierce, an expert on insect evolution and symbioses, with a focus on butterflies and ants. Naomi is not only a global authority in entomology, but also one of the kindest people I have met, and I thank her for all her guidance and support.

In 2011, I relocated to Manhattan to take up a position at Rockefeller University. The university is located a pleasant walk through Central Park away from the American Museum of Natural History, where Theodore Schneirla conducted his behavioral studies of army ants, which spanned several decades until his death in 1968. The inspiring intellectual environment at Rockefeller University has since enabled me to take my research in entirely new directions while simultaneously giving me the freedom to pursue my passion for fieldwork and natural history in complementary ways. My brilliant colleagues here are too numerous to mention by name. I am also grateful to all the members of my laboratory, past and present, who have contributed so much professional excitement and personal joy to my life. Among them, two people deserve a special thank you: Leonora Olivos-Cisneros and Stephany Valdés-Rodríguez. Leonora's friendship and expertise, both scientific and managerial, have been invaluable for all aspects of our work, and Stephany's patience and love of ants allows us to do the kind of science we are now able to do.

My work on army ants and their relatives has heavily relied on field companions and local collaborators: Patrick Kamukunji, Dino Martins, James Meitiaki, Joseph Murithi, Ivy Ng'iru, Washington Njagi, Ben Obanda, and Wilberforce Okeka in Kenya; Rene Albarán, John Lattke, Carlos Laucho, Ronald Luján, and Lenin Reyes in Venezuela; Kazuki Tsuji in Japan; Greg Zolnerowich in Kansas; and Adrián Pinto Tomás in Costa Rica. Several additional friends and colleagues have accompanied me during fieldwork over the years, including Ian Butler, Marcell Peters, and Caspar Schöning in Kenya; Philipp Hönle and Christoph von Beeren in Costa Rica; Nikolai Knoke in Venezuela; Milan Janda in Australia; and Edith Rodríguez in Venezuela, Brazil, Costa Rica, and the United States. I also thank all the fantastic staff and scientists at La Selva Biological Station for their support during our fieldwork in Costa Rica, in particular Danilo Brenes, Carlos de la Rosa, Bernal Matarrita Carranza, and Ronald Vargas. Similarly, I thank Fardosa Hassan and Cosmas Nzomo for their help in organizing a wonderful stay at Mpala Research Centre in Kenya.

Several specialists have helped identify the subjects of my photographs: Grace Barber, Isidro Chacón, Don Feener, Georg Fischer, Philipp Hönle, Milan Janda, Hans Klompen, Jack Longino, Munetoshi Maruyama, Piotr Naskrecki, Marcell Peters, Frank Rheindt, Ben Rowson, Caspar

Schöning, John Stireman, Peter Tattersfield, and Christoph von Beeren. I also thank Mark Wong for allowing me to photograph his *Neivamyrmex* fossil, and Adrian Smith for sharing his *Syscia madrensis* colony.

Vikram Chandra and Christoph von Beeren have discussed army ant biology extensively with me over the past years and provided helpful suggestions to improve the text at various stages. I thank Bettina and Hans Weidenmüller, who allowed me to spend a wonderfully quiet week at their apartment when I returned to Heidelberg in the summer of 2019 to finish a first manuscript draft. I also thank Bert Hölldobler, Chris Reid, and an anonymous reviewer for their constructive feedback on that early version. Janice Audet, Emeralde Jensen-Roberts, and Stephanie Vyce at Harvard University Press offered precious guidance and advice during the preparation of the manuscript and its figures, and Janice went to great lengths to propose detailed and thoughtful edits throughout the entire book. Kim Giambattisto and Vickie West at Westchester Publishing Services copyedited the manuscript. Marie Droual helped correct the page proofs and compose the index.

Finally, I thank Julie Pastagia, my soulmate and wife. Julie has provided many insightful and constructive comments on different aspects of the book, from the title, to the choice of images, and different text passages. During the project's early stages, we traveled to Brazil and Costa Rica together, where I took some of the pictures included in the final product. I am incredibly lucky to have you!

Daniel J. C. Kronauer
The Rockefeller University
New York, NY

index